Composting Mastery

Build Richer Soil, Grow Healthier Plants, and Turn Everyday Scraps into Garden Gold — In Any Space

Oliver Thorne

Novus Liber

First edition May 2026

Paperback ISBN 978-1-971761-08-4

Contents

PART I

THE COMPOSTING FOUNDATION

Understanding the Alchemy of Decay

Chapter 1

What Composting Really Is: The Cycle Behind the Magic

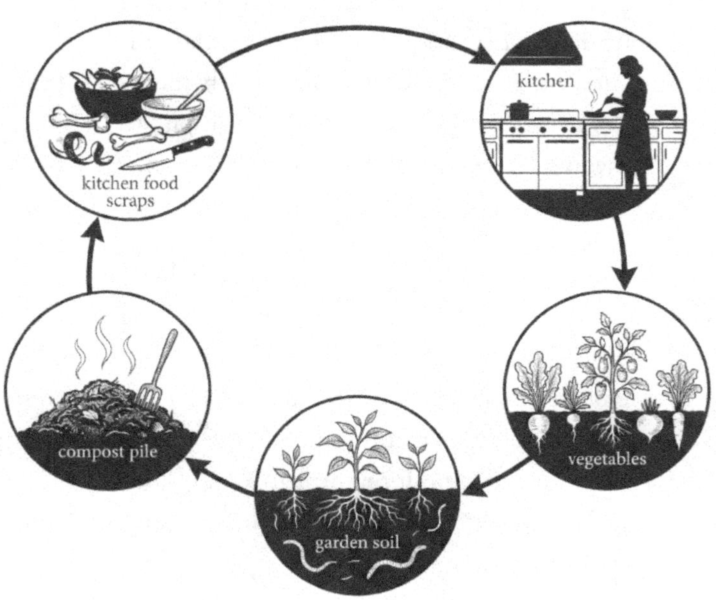

My father kept a compost heap at the back of his vegetable garden that he treated with roughly the same level of ceremony as the garden itself. He turned it every Saturday morning, checked its temperature with a long-stemmed thermometer he'd bought at a farm supply store in 1974, and talked about it as if it were a living thing — which, of course, it was. I thought he was eccentric until I had a vegetable garden of my own.

That's the moment composting stops being an abstraction and becomes a practice: when you have something to feed it and something to gain from its output. The science is not complicated — but it rewards knowing. A composter who understands why a pile heats up, why the balance of materials matters, and why some piles smell right while others go wrong will always outperform one who simply follows instructions by rote.

Definition and Scope

Composting is a controlled, accelerated version of what happens everywhere in nature when organic matter dies. Leaves fall, animals decay, wood rots — and in each case, microorganisms break down complex organic compounds into simpler ones that plants can absorb. That process happens whether you intervene or not. Composting simply manages it: gives it the right conditions, the right mix of materials, and a container or location that keeps things together long enough to produce something useful.

The difference between composting and rotting is not biological — it's managerial. Both processes involve decomposing microorganisms breaking down organic matter. In a well-managed compost pile, those microorganisms are working aerobically: with oxygen, at temperatures that pasteurize pathogens, producing CO_2 and water as byproducts. In an unmanaged rotting situation — think food in a sealed bin with no air circulation — decomposition turns anaerobic, producing methane and hydrogen sulfide. That's the difference between finished compost that smells like a forest floor and a bin you don't want to open.

This distinction matters practically. It's the difference between a pile that works quietly and efficiently and one that becomes a problem.

A Brief History

Composting has no single inventor. The practice of returning organic matter to the soil predates written history. What we do have are written records — Roman agriculturalists documenting the benefits of stacked manure and plant debris, Chinese farmers using nightsoil and organic residues in ways that kept their fields productive for thousands of years without the nutrient depletion that plagued European agriculture.

The Roman writer Columella, writing in the first century AD, described specific composting practices with a precision that would not have been out of place in a modern gardening manual. He understood, without the language of microbiology, that a managed mixture of organic materials produced better results than any single input applied raw.

The modern scientific foundation came from Albert Howard, a British agronomist working at Indore, India in the 1930s. Howard spent years studying traditional Indian composting methods and developed what became known as the Indore method: a systematic layering and turning process that produced reliable finished compost in two to three months. His 1940 book An Agricultural Testament became foundational reading for the organic farming movement that followed.

In the decades after Howard, figures like Robert Rodale and Patricia Lanza brought composting into suburban and urban consciousness. Rodale's publishing empire in Emmaus, Pennsylvania made composting accessible to generations of American gardeners who had never heard of the Indore method. Lanza's 1998 book Lasagna Gardening introduced sheet composting to readers who didn't have space or time for a traditional pile.

The line from Columella to your kitchen counter runs unbroken for two thousand years.

Composting in the Global Carbon Cycle

Carbon and nitrogen cycle through every living system on Earth. Plants absorb carbon dioxide from the atmosphere and fix it into organic compounds — sugars, starches, cellulose, lignin. When those plants die, microorganisms break the compounds back down, releasing carbon as CO_2 or, in anaerobic conditions, as methane. The nutrients stay in the soil, available for the next generation of plants.

This cycle matters to composting in two ways. First, a well-managed aerobic compost pile releases carbon dioxide — which, while a greenhouse gas, has a global warming potential roughly 25 times lower than methane over a 100-year period. Organic matter sent to a landfill, where it decomposes anaerobically, produces methane. The same kitchen scraps, composted aerobically at home, produce far less climate impact.

Second — and this is the part that takes longer to appreciate — organic matter incorporated into soil doesn't all become CO_2 immediately. A significant fraction is stabilized as humus: complex organic compounds that persist in soil for decades. Compost-amended soil is, in a measurable sense, a carbon sink. Not a large one on a global scale, but a real one. And unlike most carbon capture strategies, it produces something immediately useful.

Framing the Book

Six books follow this one. The first four build understanding and practical skill — from the science of decomposition through every major composting method to urban and indoor systems. Book V is about using what you've made: how to read your soil, apply compost where it will do most good, and integrate it with mulching, cover crops, and crop rotation. Book VI steps back from the garden entirely to look at composting in the context of climate, economics, and what comes next.

Where to start depends on where you are. Complete beginners will get the most from reading Books I and II before making any decisions about systems or materials. If you've composted before and want to expand your methods, Book III is a self-contained reference. Urban composters short on space should read Book IV alongside whatever else interests them. And if you already have a pile running and want to know exactly how to apply the output, Book V is the place to go.

Whatever your starting point, one assumption runs through every chapter: you are capable of understanding how this works, not just following a recipe. The recipe is included. But so is the reasoning.

Chapter 2

The Living Engine: Bacteria, Fungi, and the Microscopic World

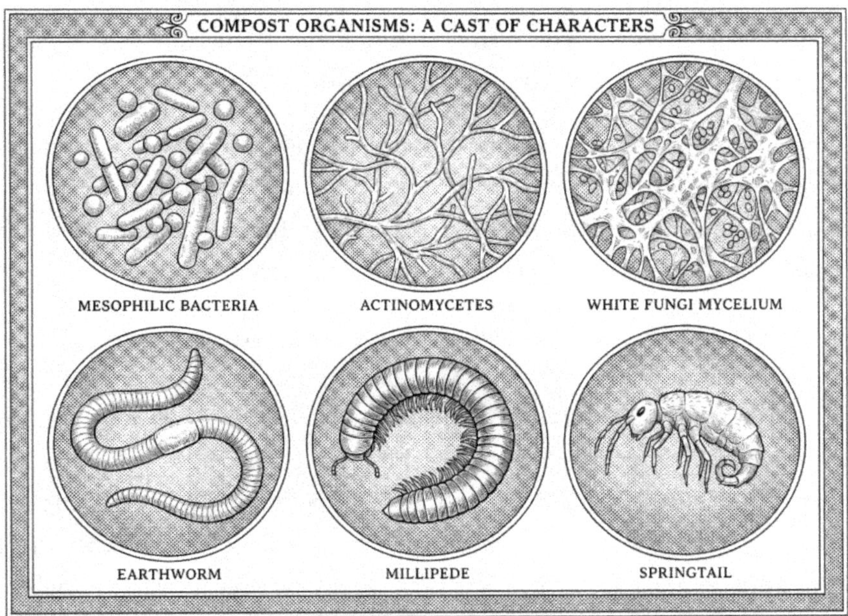

Reach into a working compost pile and you will feel heat that has no obvious source. There is no combustion, no chemical reaction you can point to — just kitchen scraps and dead leaves, slowly becoming something else. The heat is metabolic: millions of microorganisms eating, reproducing, and dying, generating warmth as a byproduct of their activity. A compost pile is not inert matter passively breaking down. It is an ecosystem — layered, competitive, and responsive to every condition you create for it.

Understanding who lives in that ecosystem, and what each group of organisms does, is the most useful thing you can learn about composting. It turns troubleshooting from guesswork into diagnosis.

Bacteria: The Primary Decomposers

Bacteria are the most numerous organisms in any compost pile and do the bulk of the decomposition work. They are also temperature-sensitive in ways that determine how a pile behaves over time.

Psychrophilic bacteria are active at low temperatures — between 28°F and 55°F (-2°C to 13°C). They are the organisms responsible for decomposition in cold compost piles and during winter months. Slow workers, but persistent.

Mesophilic bacteria take over as temperatures rise above 55°F (13°C), reaching their peak activity between 70°F and 90°F (21°C to 32°C). In a freshly built pile, these are the organisms doing the initial breakdown. Their metabolism generates heat, which eventually raises the pile's temperature beyond their preferred range.

Thermophilic bacteria are the elite decomposers of the compost world. They operate between 104°F and 170°F (40°C to 77°C), with a peak around 131°F to 160°F (55°C to 71°C). This is the hot phase — the phase that kills weed seeds and pathogens, breaks down tough materials, and produces the fastest decomposition. The first time I stuck a thermometer into a properly built pile and watched it climb to 145°F on day three, I understood why people get obsessive about this. That warmth — a pile of kitchen scraps and dead leaves generating enough heat to pasteurize itself — felt like a small miracle. It still does.

After the hot phase subsides — either because food sources have been depleted or because the pile has been turned and cooled — mesophilic bacteria return to finish the job during curing.

Actinomycetes: The Thread-Like Decomposers

Actinomycetes are bacteria that grow in thread-like filaments, resembling fungi to the naked eye. They are specialists: slow but capable of breaking down lignin, cellulose, and other tough organic compounds that other bacteria leave behind. They are the organisms responsible for the characteristic earthy, forest-floor smell of finished compost — that smell comes from a compound they produce called geosmin.

You often see them as gray or white filaments on partially composted woody material. If your pile smells right but looks like it still has work to do in the woody sections, actinomycetes are the organisms finishing that task.

Fungi and Mold: The Cellulose and Lignin Specialists

Fungi are the kingdom most capable of breaking down lignin — the complex polymer that gives wood its structural strength and that most other organisms cannot digest. Without fungi, woody materials would persist in a compost pile for years.

White patches on compost or on the inside surfaces of a bin are mycelium — fungal threads doing their job. Blue or green patches are mold species, also decomposers. Neither is cause for concern. They are indicators of active biological work, not contamination.

Fungi thrive in cooler, less-disturbed piles. Frequent turning disrupts fungal networks and tends to favor bacteria. This is one reason why cold, undisturbed compost piles often produce a different type of output than frequently turned hot piles — a distinction that has real consequences for how you use the finished product.

Fungal vs. Bacterial Compost: A Crucial Distinction

The balance of fungi and bacteria in a finished compost is not just a biological curiosity — it determines which plants that compost will benefit most.

Bacterial-dominated compost is produced by hot piles that are turned frequently. The heat and disturbance favor bacteria over fungi, producing a nutrient-dense, rapidly available compost. This type is ideal for vegetables, annual flowers, and other crops that prefer high available nitrogen and rapid nutrient cycling.

Fungal-dominated compost comes from cool, undisturbed piles with a high proportion of woody material. Fungi thrive without being disturbed. This type of compost inoculates soil with fungal networks that form symbiotic relationships with plant roots — relationships that are essential to trees, shrubs, perennial vegetables, and most established garden plants.

The practical implication: if you are amending a vegetable bed, a hot-turned compost is your best input. If you are planting a fruit tree or enriching an established shrub border, a cool fungal compost applied as a mulch — not incorporated — is the more appropriate choice.

You can intentionally steer your pile toward one type or the other. More turning, higher nitrogen, and a finer particle size push it toward bacterial dominance. Less turning, more woody material, and larger particle sizes favor fungi.

Protozoa and Larger Organisms

Protozoa are single-celled organisms that feed on bacteria. In doing so, they release nutrients — particularly nitrogen — that bacteria had incorporated into their own cells. Protozoa are part of what makes finished compost a more complete fertilizer than the sum of its ingredients: they are a link in the nutrient mineralization chain.

At the larger end of the biological spectrum, compost piles host a visible ecosystem. Earthworms — the same species that will populate compost-amended garden soil — process organic matter and leave behind castings that are nutritionally superior to the surrounding compost. Beetles and their larvae shred tough materials into smaller particles that bacteria can access. Millipedes, centipedes, mites, and springtails each occupy specific roles in the decomposition chain.

The diversity of this community matters. A pile with high biological diversity — driven by varied inputs, appropriate moisture, and adequate aeration — decomposes faster and produces a more nutritionally complete output than a pile dominated by a single class of organisms.

The Soil Microbiome and Human Health

This is an area of emerging science, and the appropriate response to emerging science is informed interest, not evangelism. What the research suggests, without overstating it, is that the diversity of microorganisms in soil influences the nutritional diversity of the food grown in it.

Plants interact with soil microorganisms through their root systems — releasing compounds that attract specific microbial communities, and receiving in return nutrients that microbial activity makes available. The nutritional profile of a crop is not determined solely by the N-P-K content of its soil. It is shaped by the microbial ecosystem around its roots.

Research linking soil microbiome health to human gut microbiome health is at an early stage, but it points in a consistent direction: diverse soil produces more nutritionally varied food, and nutritional diversity correlates with gut microbial diversity. The gut-soil connection is not established fact, but it is serious science.

For the composting gardener, the practical takeaway is this: adding high-quality, biologically diverse compost to soil is not only a fertility strategy. It may also be a food quality strategy. That is a claim I make carefully, and with the acknowledgment that the science is still developing. It is, however, one of the more compelling personal arguments for composting.

Chapter 3

The Chemistry of Compost: Carbon, Nitrogen, and the Art of Balance

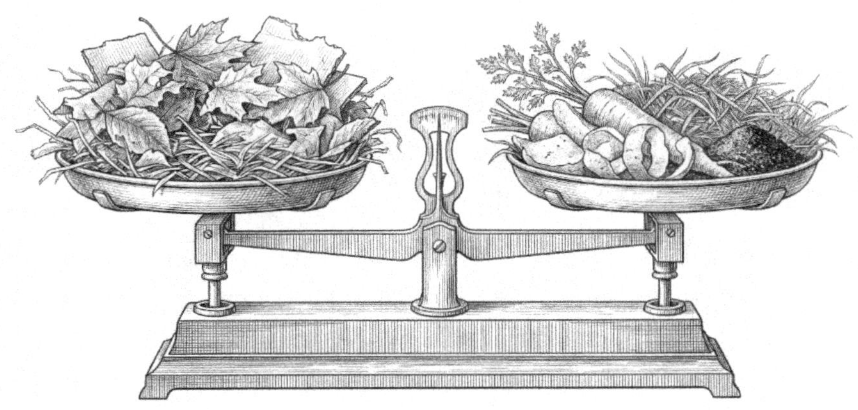

S queeze a fistful of your compost. If water streams out freely, the pile is too wet. If the material falls apart into dry powder the moment you release pressure, it's too dry. The right moisture level is the wrung-out sponge: damp throughout, with only a few drops falling when you squeeze firmly. That physical test tells you more in two seconds than a moisture meter will tell you in ten minutes.

Chemistry governs composting. Not the chemistry of a lab — the chemistry of living systems, where the relevant numbers are ratios, not formulas, and where small adjustments produce large effects. This chapter covers the core chemistry you need: the C:N ratio, temperature, moisture, aeration, and pH. Understand these, and you can diagnose any problem a pile presents.

The C:N Ratio: Explained Once, Completely

Carbon and nitrogen are the two most important elements in composting, and the balance between them determines everything from how fast your pile decomposes to what it smells like.

To a decomposing microorganism, carbon is energy and nitrogen is protein. Microbes need both. They consume carbon as fuel for their metabolism and use nitrogen to build new cell proteins as they reproduce. The ratio at which they consume these elements is approximately 25 parts carbon to 1 part nitrogen by mass — written as 25:1.

When your pile has a ratio significantly above 30:1 — too much carbon relative to nitrogen — microorganisms work slowly. There isn't enough nitrogen to support rapid reproduction. The pile may feel dry, look brown and inactive, and produce little heat. It will eventually decompose, but slowly.

When the ratio falls significantly below 20:1 — too much nitrogen relative to carbon — microorganisms reproduce rapidly but produce ammonia as a byproduct. That ammonia is the source of the sharp, eye-watering smell that indicates a pile with too many greens and not enough browns. You are also losing nitrogen to the atmosphere, which is a waste of a valuable input.

The target is 25 to 30:1. In practice, this translates roughly to one part green material by volume for every two to three parts brown material. This is not a precise formula — it's a working estimate that you adjust based on what you're adding and how your pile responds.

C:N Values by Material

The following ratios are averages. Individual values vary by species, age, and condition of the material.

Browns (high carbon): Dried leaves run roughly 60:1. Straw is around 80:1. Cardboard ranges from 300:1 to 400:1. Sawdust can reach 500:1. Wood chips are similar to sawdust. These materials add carbon to your pile but contribute almost no nitrogen on their own.

Greens (higher nitrogen): Grass clippings are approximately 20:1. Vegetable scraps from the kitchen average 15:1. Coffee grounds, despite their brown color, run about 20:1 and are considered a green. Fresh manure ranges from 7:1 (chicken) to 25:1 (horse).

Animal manures: Chicken manure is nitrogen-rich at around 7:1 — it acts as a powerful activator but can push a pile toward the ammonia-producing side if overdone. Horse manure at 25:1 is almost perfectly balanced on its own. Cow manure at 20:1 is slightly nitrogen-rich. Rabbit manure at 13:1 is among the most nitrogen-dense of the common options.

How to use these numbers: before you build a pile, estimate the ratio of your available materials. If you have a large volume of wood chips (500:1) and kitchen scraps (15:1), you need significantly more scraps by volume to bring the ratio into range. If you have straw (80:1) and fresh grass clippings (20:1), a roughly 2:1 mix by volume gets you close.

Temperature: The Thermometer of Microbial Health

A compost pile that is working correctly heats up. This is not optional — it is the observable evidence that microbial activity is occurring at the rate required to decompose materials efficiently and kill pathogens.

The three temperature phases follow a predictable arc. In the mesophilic phase, which lasts from a few days to a week, temperatures rise from ambient to around 100°F (38°C) as mesophilic bacteria colonize the new material. In the thermophilic phase, temperatures climb to between 131°F and 160°F (55°C to 71°C). This is where the most active decomposition occurs and where weed seeds and human pathogens are killed. After turning and re-activating the pile, a second thermophilic phase may occur. The curing phase that follows is a return to cooler mesophilic activity as the most available materials have been consumed.

The USDA standard for pathogen destruction requires temperatures of at least 131°F (55°C) maintained for a minimum of three consecutive days. For a pile to reach this temperature, it must be large enough to insulate its own heat — typically at least one cubic meter (roughly 3 feet by 3 feet by 3 feet). Below this size, heat dissipates faster than it can be generated.

Temperatures above 170°F (77°C) are a danger zone. At these temperatures, even thermophilic bacteria begin to die, and you can permanently damage the microbial community in your pile. If your thermometer reads above 160°F, turn the pile to cool it and increase aeration.

Read your thermometer as a diagnostic tool, not just a milestone. A temperature that is climbing tells you the pile is active. A plateau followed by a steady decline tells you the hot phase is ending and the pile may benefit from turning. A temperature that never rises above ambient tells you something is missing — usually insufficient nitrogen, moisture, or mass.

Moisture: The Medium of Life

Microorganisms live and function in water. They need moisture to move through the pile, to access food sources, and to excrete waste products. A pile that dries out simply stops working. A pile that becomes waterlogged goes anaerobic, because water displaces the air that aerobic organisms need.

The target moisture level is 50 to 60% by weight. In practical terms, this means materials that feel damp throughout but do not drip freely when handled. The squeeze test remains the most reliable field method: take a large handful of material from the interior of the pile and compress it firmly. A few drops of water should emerge. If the material streams water, it's too wet. If nothing happens and the material crumbles, it's too dry.

Moisture meters exist, and they're useful for gardeners who want precision. But the squeeze test has the advantage of requiring nothing more than your hand, and it works accurately enough that most experienced composters never use anything else.

Aeration: Oxygen as the Engine

Aerobic decomposition — the kind we want in a compost pile — requires oxygen. Oxygen allows thermophilic bacteria to operate at peak metabolism, producing heat and breaking down materials quickly. Remove oxygen from the equation and decomposition continues, but anaerobically: slowly, incompletely, and with unpleasant byproducts.

Aeration in a compost pile happens in two ways: passive air movement through the pile's structure, and active turning. Passive aeration depends on the pile's porosity — its ability to allow air to move through it. Coarse, varied materials maintain better structure and allow better air movement than fine, uniform materials that compact together. This is one reason shredding materials into small particles, while it speeds decomposition by increasing surface area, needs to be balanced against the compaction it creates.

Turning the pile is the most effective way to restore oxygen levels after they have been depleted. It also moves outer material to the center (where the heat is) and breaks up any anaerobic pockets. How frequently to turn depends on your goals: more frequent turning produces faster decomposition but more labor. Cold composting involves little or no turning and simply requires more time.

pH Dynamics Throughout the Composting Process

A fresh compost pile is typically acidic. As organic materials begin to decompose, organic acids — acetic acid, lactic acid — are produced as intermediate byproducts, driving pH down to around 5 or 6. You may notice a sour smell in the early stages of a pile's life. This is normal.

As the thermophilic phase proceeds, pH rises. By the time a pile reaches the active hot phase, it has typically moved toward neutral or slightly alkaline — around 7 to 7.5. Finished compost is nearly neutral, which is close to the optimal pH range for most garden plants.

Significant deviations from this pattern are diagnostic signals. A pile that remains very acidic throughout may have too much fruit waste or conifer needles. A pile that becomes very alkaline may have too much wood ash. Neither extreme prevents decomposition, but both slow it and can affect the quality of the finished product.

Most composters never need to actively adjust pH. The process self-corrects if the C:N ratio and moisture are in range. If you suspect pH is an issue, a standard soil pH test kit works on compost material as well as soil.

Chapter 4

Why Compost? The Case for Greens, Browns, and a Better Planet

Approximately 30% of what the average American household throws away could be composted. Not could be recycled in some industrial process, not could theoretically be diverted — could be processed right now, at home, with no special equipment, into something directly useful. That number is not an estimate or a projection. It is based on waste composition studies conducted across dozens of municipalities over decades. The material is there. The question is what happens to it.

This chapter makes the case for composting with data rather than ideology. The environmental argument, the economic argument, and the personal argument are each real. Each deserves accurate numbers rather than impressionistic claims.

Environmental Benefits: With Data

The United States generates approximately 80 million tons of organic waste annually. Of this, roughly 35 million tons ends up in municipal landfills, where it decomposes anaerobically and produces methane. Landfill methane from organic waste accounts for approximately 20% of US methane emissions. Methane has a global warming potential 80 times that of CO_2 over a 20-year period.

Composting that same material aerobically produces CO_2 instead — still a greenhouse gas, but one with a fraction of the climate impact. And beyond the emissions prevented, the compost produced improves soil in ways that have their own climate benefits: compost-amended soil stores carbon, reduces erosion, and decreases irrigation needs.

Water conservation is one of the least-discussed environmental benefits of composting. Soil amended with mature compost can reduce irrigation requirements by 20 to 30%. Compost particles absorb and hold water, releasing it slowly as plants need it. In regions facing water stress, that 20 to 30% is not a rounding error — it translates directly to lower water bills and less pressure on local supply.

Biodiversity in compost-amended soil runs measurably higher than in depleted or synthetically fertilized soil. Studies have documented up to six times more species of soil organisms in compost-rich plots compared to conventionally managed ones. That diversity is not just ecologically interesting — it is functionally important for plant health, disease suppression, and nutrient cycling.

Economic Benefits: With Data

A well-managed home compost pile producing mature compost consistently can replace between \$150 and \$250 per year in purchased organic fertilizers and soil amendments, depending on what you're growing and where you live. That range is based on retail prices for comparable volumes of organic compost, vermicast, and slow-release fertilizers — products that a home compost operation produces at no cost beyond time.

Water bill savings are harder to calculate precisely because they depend on local water prices, climate, and how extensively you apply compost. But municipalities that have tracked irrigation use in community gardens before and after compost amendment have consistently documented 20 to 30% reductions in water usage. At \$0.01 to \$0.02 per gallon — typical municipal water rates — that adds up over a growing season.

At the municipal level, the economics are compelling. Organic material diverted from landfills reduces disposal costs — landfill gate fees run between \$40 and \$70 per ton in most US municipalities. Programs that capture organics for composting instead consistently show lower per-ton processing costs and an additional revenue stream from finished compost sales.

Long-term soil capital is the economic argument that is hardest to quantify but most significant over time. Healthy topsoil — the top six to eight inches of soil with high organic matter content, strong microbial activity, and good structure — takes decades to build naturally and years to build with consistent compost amendment. Studies examining the economic value of topsoil put it at anywhere from \$1,000 to \$100,000 per acre, depending on agricultural productivity. Every cubic yard of compost you incorporate is an investment in that capital.

Personal Benefits

The environmental and economic case has been made above, with numbers. But that's not why most people continue composting once they've started. They continue because it works — visibly, tangibly, in their own gardens.

There is a specific satisfaction in closing a loop that modern life has trained us to leave open. Food scraps go in one end; improved soil comes out the other. What was garbage becomes a resource — and that reorganization is practical, not poetic.

For gardeners who grow food, the link between healthy compost-amended soil and the quality of what you harvest is direct enough to taste. Tomatoes grown in well-composted soil consistently outperform those grown in depleted or synthetically fertilized soil in flavor assessments. This is partly nutrition, partly water content, partly microbial complexity. The point is that the quality difference is real and perceptible.

There are also psychological benefits to tending a compost pile that are worth acknowledging without exaggerating. Observational care for a living system — checking its temperature, adjusting its moisture, watching materials transform over weeks — engages the same cognitive patterns as any other form of careful attention to natural processes. Whether that qualifies as mindfulness or simply as a satisfying project is for you to decide. The people who compost for years do not do it primarily out of environmental conviction. They do it because it works, and because it connects them to the soil they garden in.

The Bigger Picture

No individual composter is going to solve the global organic waste problem. That should be said clearly, without apology. What individual composting does is remove your share of the problem from the system — and, increasingly, contribute to a shift in cultural expectation about what belongs in a landfill.

The cities and regions that have made the most progress on organic waste diversion — San Francisco, Seattle, Vermont — did so because of policy, infrastructure, and public adoption working together. Individual action created the constituency that made the policy possible. That is how these transitions work in practice: not top-down or bottom-up exclusively, but both simultaneously.

Composting sits within the broader framework of regenerative agriculture — a collection of practices aimed at restoring soil health, reducing agricultural chemical dependency, and building farming systems that improve over time rather than degrade. Home composting is the smallest scale at which regenerative principles operate. The principles scale upward from there.

Chapter 5

The Chemistry of Finished Compost: Nutrients, Numbers, and Comparisons

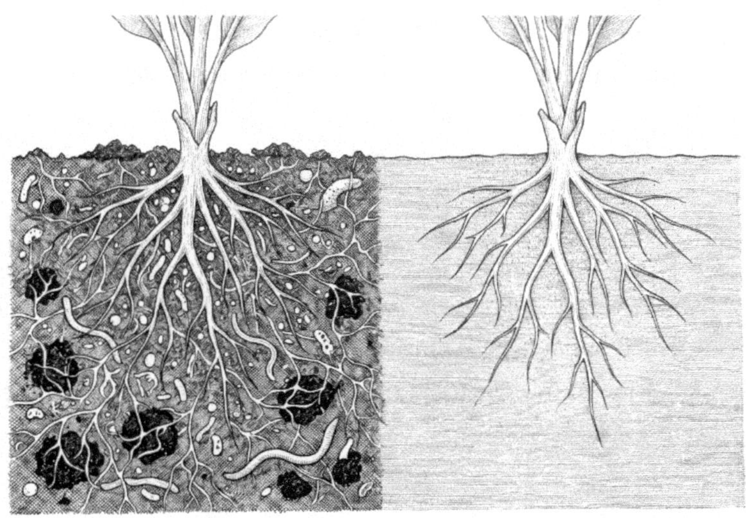

The bag test is simple. Take a sample of what you think is finished compost, seal it in a clear plastic bag, and leave it in a warm spot for three days. When you open it, smell it. If the smell is earthy and clean — that forest-floor quality — the compost has cured properly. If there's sourness, ammonia, or sulfur, decomposition is not complete. Three days in a sealed bag will tell you more reliably than any visual inspection.

What follows is the chemistry of finished compost — what it contains, how those contents compare to what synthetic fertilizers deliver, and why the inputs you chose months ago are now reflected in what you have in your hands.

What Mature Compost Contains

Home compost typically analyzes at approximately 1% nitrogen, 0.5% phosphorus, and 0.5% potassium by weight — written as 1-0.5-0.5 in the N-P-K notation used for fertilizer labeling. These are average figures; actual values vary considerably depending on input materials, composting method, and curing duration.

Compared to synthetic fertilizers — which might label as 10-10-10 or 20-10-5 — these numbers look modest. The comparison is misleading. Synthetic fertilizer values represent immediately available nutrients. Compost values represent total nutrients, including a significant fraction that will be released slowly over weeks and months as microbial activity continues to break down organic matter in the soil. That slow release matches plant demand more accurately and produces less nutrient runoff.

Beyond N-P-K, mature compost contains a full spectrum of micronutrients — iron, manganese, zinc, copper, boron — in forms that plants can access. Synthetically fertilized soil frequently develops micronutrient deficiencies over time because synthetic applications don't replace what crops remove. Compost applications replenish the full spectrum.

Humic and fulvic acids, produced during the curing phase of composting, are organic compounds that play a specific role in soil chemistry. They improve the ability of soil particles to hold cations — positively charged nutrient ions — and make those nutrients more accessible to plant roots. This is the mechanism behind what gardeners describe as compost improving the "structure" of soil: it's not just physical, it's chemical.

A gram of finished compost from a well-managed pile contains somewhere between 100 million and 1 billion bacteria, along with significant populations of fungi, actinomycetes, protozoa, and nematodes. Those organisms don't just sit dormant — they continue working in the soil, cycling nutrients, building structure, and suppressing some plant diseases. This is the component of compost value that no synthetic product replicates.

Compost vs. Synthetic Fertilizers

Synthetic fertilizers provide nutrients in immediately water-soluble forms. Plants can take them up right away, which is useful when a crop needs a quick boost. The problem is that what plants don't take up immediately leaches through the soil profile and enters groundwater or runs off in irrigation water. Nutrient pollution of waterways — algal blooms, hypoxic zones — is substantially driven by synthetic fertilizer runoff from agricultural land.

Over years of synthetic fertilizer use without organic matter additions, soil structure degrades. The microbial populations that build soil structure, cycle nutrients, and suppress disease are not fed by synthetic fertilizers. Soil becomes compacted, nutrient-dependent on external inputs, and progressively less capable of supporting crops without increasing synthetic applications. That trajectory is documented across decades of agricultural research — not a warning, a record.

Compost used as the primary soil amendment does not produce these effects. Nutrients are released slowly, losses through leaching are minimal, and soil structure improves over time rather than degrading. The case for using compost as your primary fertility input and synthetic fertilizers only as targeted supplements — if at all — is well supported by long-term research.

How Input Materials Shape Output Quality

Compost made primarily from food waste — kitchen scraps, coffee grounds, fruit and vegetable residues — tends to be higher in nitrogen and phosphorus than yard-waste-dominated compost. Food waste is nutritionally dense material, and the compost reflects that.

Compost dominated by yard waste — leaves, grass clippings, woody prunings — tends to have a higher carbon content, lower available nitrogen, and more humic compounds from the lignin-rich materials. It is often better for building soil structure than for delivering immediate fertility.

You can enrich compost for specific purposes before application. Adding a finished vermicast layer to a basic compost significantly raises its microbial density and nutrient availability. Incorporating seaweed meal adds trace minerals and growth-stimulating compounds. Rock dust, blended into compost before application, provides a slow-release source of silica and a range of trace minerals.

Recognizing High-Quality Finished Compost

Good finished compost is dark brown to black. The color comes from humic compounds formed during curing. If your compost is pale brown or light tan, it likely hasn't finished curing, even if the original materials are no longer recognizable.

The texture should be crumbly and loose — not compacted into a dense mass and not powdery dry. High-quality compost feels like the best garden soil you've ever worked with: soft, porous, and uniformly moist.

No original materials should be recognizable. If you can still see vegetable scraps or leaf fragments, the compost isn't finished. Screen it if you want to use it now — the screened-out material goes back into the next pile — or leave it for another few weeks.

The smell should be the most reliable indicator. Earthy, complex, slightly sweet, reminiscent of a forest floor after rain. Someone once asked me at a workshop what finished compost should smell like, and I said: October morning in the woods. They nodded. If the smell has any sharpness — ammonia, sourness, sulfur — it's not done.

Immature compost applied to a garden creates two problems. First, it continues decomposing in the soil, drawing nitrogen from the surrounding environment — a phenomenon called nitrogen theft — which can actually deprive plants of nitrogen during a critical growth period. Second, some intermediate decomposition products — particularly organic acids — are phytotoxic: directly harmful to plant roots and seeds. A one-off mistake with established plants is unlikely to be serious. Applied in a seed bed or planting hole with direct root contact, though, immature compost can kill seedlings or significantly slow germination.

Wait until it's ready. The bag test works. The nose test works. Don't rush the last step.

YOUR FIRST COMPOST PILE

A Practical Guide
from Zero to Black Gold

Chapter 6

Choosing Your System: Bins, Tumblers, and Open Piles

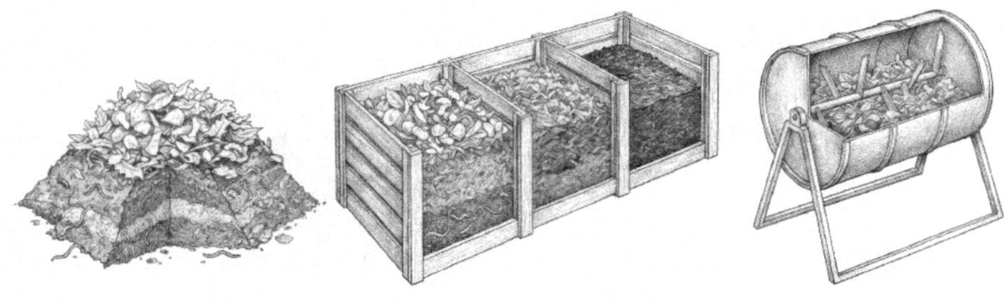

My neighbor — a sensible, pragmatic man who grew up on a farm — watched me set up my first tumbler with undisguised skepticism. "You're paying for a machine to do what a pile in the corner does for free." He wasn't wrong, exactly. The tumbler did compost faster. But I stopped using it within two years because turning it every three days in a wet winter is less pleasant than it sounds in the brochure.

The right composting system is the one you will actually use consistently. That sounds obvious, but the composting equipment market is full of products designed for an idealized composter who has more time, space, and enthusiasm than most people can sustain. This chapter helps you find the system that matches your real life, not the aspirational version.

The Decision Framework

Four variables determine the right system: available space, weekly waste volume, time budget, and climate.

Available space sets the upper limit on what's possible. An apartment kitchen with no outdoor access eliminates outdoor piles and requires an indoor system. A suburban backyard with a 20-foot setback requirement opens most options. A rural property with half an acre removes nearly all constraints.

Weekly waste volume is often underestimated. The average American household generates approximately 4 to 6 pounds of compostable kitchen waste per week, plus variable quantities of yard waste. A single-person household might generate 2 to 3 pounds. A family of four with a large garden can produce 15 to 20 pounds in peak season. The system you choose needs to handle your actual volume, not your average estimate.

Time budget is honest and important. Some systems require regular turning — every three to five days in hot composting, weekly in a tumbler. Others require almost nothing but occasional moisture checks. If you work fifty hours a week and travel frequently, a high-maintenance system won't survive your first busy month. If you're a retired gardener who enjoys regular outdoor tasks, a three-bin turning system is a pleasure.

Climate shapes system performance more than most guides acknowledge. In a cold northern winter, outdoor open piles go dormant. Tumblers freeze. A worm bin in an uninsulated garage can kill your entire worm population below 40°F. If you compost year-round in a cold climate, your system design needs to account for winter.

Open Pile / Heap

The open pile is composting in its oldest and simplest form: organic materials stacked in a corner of the yard, ideally in a spot with good drainage and reasonable proximity to both the kitchen and the garden. No container required, no purchase necessary. The minimum effective size is roughly three feet by three feet by three feet — approximately one cubic meter. Below this size, a pile cannot retain enough heat to become thermophilic, and decomposition proceeds slowly regardless of how well-balanced the materials are.

An open pile has unlimited theoretical capacity — you simply expand it as inputs accumulate. It integrates naturally into the garden ecosystem: earthworms, beetles, and other soil organisms move freely in and out. Setup costs nothing.

What it doesn't offer is containment. An open pile is exposed to rain, which can waterlog it, and to scavenging animals drawn to food scraps. It occupies visual space. Without regular management it drifts toward slow, patchy decomposition rather than the directed process described in Chapter 3.

Stationary Bins

A stationary bin constrains and organizes the composting process without significantly changing its biology. Materials go in at the top; finished compost, eventually, comes out at the bottom or through a side access door. The bin holds material together, reduces visual clutter, offers some protection from large scavengers, and can improve moisture retention.

Purchased plastic bins are the most common option and work adequately for most situations. They are inexpensive — often subsidized by municipal programs — easy to set up, and durable. Their main limitations are fixed volume and, in some designs, poor aeration at the center of the pile. Look for models with ventilation slots on all sides, not just the lid.

DIY options range from simple wire mesh cylinders to elaborate multi-chamber wooden systems. A cylinder of hardware cloth or chicken wire, three feet in diameter and three feet tall, makes a functional open-sided bin that allows excellent aeration. Wooden pallet bins — four pallets wired together into a square — are a reliable, free option in areas where pallets are available.

The three-bin turning system is the gold standard for high-output home composting. Bin one holds fresh materials as they accumulate. When full, the pile is turned into bin two, where active hot composting occurs. When that pile has finished its thermophilic phase, it moves into bin three for curing. The result is a continuous production system: bin three is always ready to harvest as bin one fills. Building this system from wooden boards or pallets takes a weekend and several decades of use.

Tumblers

A tumbler is a sealed container mounted on a frame so it can be rotated, either by cranking a handle or by rolling it on its supports. The rotation mixes materials and aerates the pile simultaneously, making frequent turning much easier than fork-turning an open pile.

They work — particularly for kitchen waste where pest exclusion is important, and for gardeners who want speed and find the turning process physically difficult. The sealed design keeps rodents and raccoons out. The aeration from regular rotation speeds decomposition compared to a neglected open pile.

Volume is the first constraint: tumblers cap out at 50 to 110 gallons, which fills faster than most gardeners expect once autumn leaves arrive. They are also an upfront expense, and they require consistent turning — a tumbler loaded and forgotten becomes a sealed anaerobic container, not a composting system.

When shopping for a tumbler, look for models with two independent chambers (so you can add fresh material to one while the other cures), substantial aeration holes rather than just a few slots, and a frame that holds the drum at a height that allows easy loading and a comfortable turning action.

Electric Countertop Composters

These are a relatively recent product category, and they deserve an honest assessment separate from the marketing claims. A countertop electric composter processes food waste using heat, aeration, and grinding into a dehydrated, reduced-volume material — which is closer to pre-compost than to the finished product described in Book I Chapter 5. It still needs soil contact to complete its breakdown.

The output does have practical applications. It breaks down faster than raw food scraps when added to a pile or soil, it's odor-neutral, and it reduces volume considerably. For apartment dwellers without outdoor access who want to divert food waste from landfill and have a community garden or planter boxes to deposit it in, an electric composter fills a specific gap.

Running one costs roughly 1 kilowatt-hour per cycle, and the machines themselves range between $300 and $500 — significant compared to free or subsidized alternatives. Use one if it fits your situation; just don't mistake its output for finished compost.

Choosing a Location

A compost system's location affects how consistently you use it more than any other design factor.

Convenience to the kitchen matters more than aesthetics. A bin that requires a long walk across the yard in winter rain gets ignored. Ideally, your composting area is within 50 to 100 feet of your kitchen door. If that's not possible, a countertop collector — a small sealed container where kitchen scraps accumulate for two to three days — bridges the gap.

Drainage is essential. A pile sitting in standing water goes anaerobic and produces the swamp-like odors that make composting unpopular with neighbors. Choose a spot with naturally good drainage, or elevate the pile base with a layer of coarse wood chips.

Partial shade is preferable to full sun in hot climates. Direct summer sun dries a pile rapidly, requiring constant moisture adjustment. In cool climates, a sunny spot helps maintain temperatures.

Distance from property lines and water sources is both practical and sometimes legally required. Check your local codes before placing a pile. Most municipalities require a minimum setback from property lines, typically five to ten feet, and many prohibit composting within a certain distance of a well or surface water.

Chapter 7
What Goes In: The Definitive Materials Guide

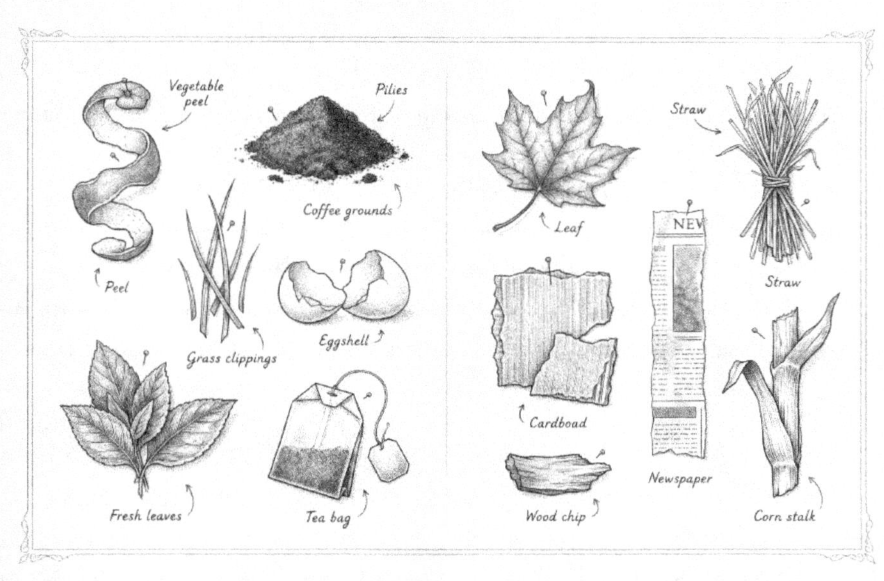

The first responders to any new compost pile are bacteria you'll never see — and they're already in your kitchen scraps, on the surface of dried leaves, in the garden soil you toss in to inoculate a new pile. You don't need to add anything special to start a pile. The organisms arrive on their own. What you control is the environment you create for them.

That environment begins with materials. What you add determines the C:N ratio, the moisture level, the aeration of the pile, and ultimately the quality of the compost you produce. This chapter is the definitive reference for what goes in and why —

including materials that are often treated as simple rules ("no meat") but deserve more nuanced answers.

Greens: Nitrogen-Rich Materials

"Greens" is the composting term for nitrogen-rich materials. They are not always green in color — coffee grounds, fresh manure, and overripe fruit are all greens in the composting sense. What defines them is their relatively low carbon-to-nitrogen ratio: typically under 30:1, and often much lower.

Vegetable and fruit scraps from the kitchen are the most common green input for most home composters. Peels, cores, rinds, overripe produce, wilted salad greens — all of these decompose readily and contribute nitrogen and moisture to the pile. There is no need to chop them, though smaller pieces do decompose faster.

Coffee grounds are among the best composting inputs available. Despite their brown color, they run approximately 20:1 C:N ratio, act as a mild nitrogen source, and are already partially broken down by the brewing process. Tea bags work similarly, though some bags contain synthetic fibers that won't decompose — if your tea bags have a smooth finish rather than a papery one, remove the bag and add only the leaves.

Fresh grass clippings are nitrogen-dense at approximately 20:1 but require attention. A thick layer of fresh clippings compacts into a dense, anaerobic mat that stops air movement and creates exactly the conditions you are trying to avoid. Add clippings in thin layers of no more than two to three inches, alternating with brown materials, or mix them into the pile rather than spreading them as a distinct layer.

Fresh plant trimmings from the garden — annuals, herbaceous perennials, vegetable tops — are excellent greens. Avoid any plant material that shows signs of disease (until you are confident your pile reaches pathogen-kill temperatures), or that you know to be particularly invasive if allowed to propagate from seeds.

Browns: Carbon-Rich Materials

Browns are carbon-rich materials: dried, woody, or papery. They are what slow a pile down when there are too many of them and what correct an over-nitrogenous, ammonia-smelling pile when you add them in the right quantity.

Dried autumn leaves are the archetypal brown material, and in many parts of the temperate world, they arrive in quantities that dwarf any other composting input. At approximately 60:1 C:N ratio, they are good but not extreme. Shredding or mowing over them before adding to the pile significantly speeds their decomposition — whole leaves tend to mat together and slow air movement.

Cardboard is excellent and readily available. Corrugated cardboard at 300 to 400:1 C:N is very high in carbon, so use it in combination with higher-nitrogen inputs. Remove any tape or staples, tear it into pieces, and moisten it before adding. Cardboard also works as the base layer of a lasagna garden or as sheet mulch — it suppresses weeds while decomposing in place.

Shredded newspaper and uncoated paper work similarly to cardboard. Avoid glossy magazine paper — the coatings resist decomposition and may contain heavy metals in the ink.

Wood chips and sawdust are extremely high in carbon — 400 to 500:1. They are best used in small quantities as part of a mixed pile, or as the base material in a hugelkultur system. Pure sawdust in large quantities will tie up nitrogen for months as it decomposes. Sawdust from chemically treated wood should never be used.

Animal Manures: A Comprehensive Guide

Manures are among the most powerful composting inputs available, and they are surrounded by more confusion than almost any other material category. The essential distinction is between hot manures and cold manures, and the critical concern is pathogen management.

Chicken manure is the most nitrogen-rich of the common options, running approximately 7:1 C:N ratio. Fresh chicken manure is potent enough to burn plant roots directly and should never be applied raw to growing beds. In a compost pile, it acts as a powerful activator — a relatively small addition can jumpstart a sluggish pile. Aged or composted chicken manure is safer for direct soil application and widely available as a commercial product.

Horse manure at approximately 25:1 is nearly self-composting — add it in large quantities and it will heat up on its own. It is commonly available from stables, often free, and already includes bedding material that provides carbon. The main concern with horse manure is weed seeds: horses don't digest seeds as thoroughly as some other animals, and hay-fed horses may produce manure loaded with viable weed seeds. Hot composting to pathogen-kill temperatures eliminates this concern.

Cow manure at approximately 20:1 is well-balanced and widely used. It composts reliably and is generally available commercially as bagged dried product. Fresh cow manure from pasture-raised cattle tends to contain fewer pathogens than manure from confined feeding operations, but all fresh manure should be composted before use on food crops.

Rabbit manure at approximately 13:1 is cold manure — it does not burn plants and can technically be applied directly to beds. Composting it first still improves its effectiveness by increasing microbial activity and improving its texture.

Goat and sheep manure, pigeon manure, and other poultry manures (turkey, duck, geese) all compost well. Their specific C:N ratios vary, but all are on the nitrogen-rich end of the spectrum.

Manures to avoid in standard home composting: dog waste, cat waste, and pig waste all carry human-transmissible pathogens including Toxocara, Cryptosporidium, and various Salmonella strains. Dedicated pathogen-destruction systems with consistent high-temperature monitoring exist for these materials, but they are not appropriate for standard home piles or use on food crops.

Special-Case Materials

Eggshells are a valuable calcium source. They decompose slowly — crushed shells may persist for a year in a cool pile — so crush them before adding, and expect them to continue breaking down in the soil after the compost has been applied. They do not contribute meaningfully to the C:N ratio.

Hair and fur from your hairbrush or your pet's grooming session are legitimate composting inputs. At approximately 6:1 C:N ratio, they are nitrogen-rich, though they decompose slowly due to their structure. Small quantities at a time.

Natural fiber fabric — 100% cotton, 100% wool, linen, hemp — composts, though slowly. Blended fabrics with synthetic content do not compost and leave synthetic residue. If you're not certain of the fiber content, don't add it.

Feathers compost but are very slow. Use small quantities.

Wood ash from an unpainted wood fire adds calcium and potassium. Use in moderation — no more than a light dusting every few months. Wood ash is alkaline and can raise pile pH above the comfort range if overused.

Seaweed and kelp are excellent inputs in coastal areas where they are available. They are nutrient-rich, break down quickly, and add trace minerals not present in most terrestrial organic materials. Rinse before adding to reduce salt content, which can inhibit microbial activity.

What NOT to Compost — and Why

Meat, fish, dairy products, and cooked food containing oils should not go into a standard home compost pile. The reason is not that they don't decompose — they do — but that their decomposition attracts specific pests. Rats are drawn to meat and dairy odors even through the walls of a closed bin. Flies and their larvae (maggots) thrive in protein-rich, moist environments. The exception is bokashi fermentation, which handles these materials specifically and is covered in Book III Chapter 4.

Diseased plants are a calculated risk. The thermophilic phase of a hot compost pile reliably kills most plant pathogens. But the key word is reliably: a pile that is not consistently reaching 131°F throughout may not fully kill diseases like tomato blight or clubroot. Until you are confident your pile is reaching and maintaining pathogen-kill temperatures, compost diseased plant material separately or discard it.

Weed seeds present the same risk. A pile that consistently reaches hot-phase temperatures kills seeds. A cold pile that never heats up does not, and the finished compost can become a vehicle for weed dispersal. Know which composting method you're using and adjust your inputs accordingly.

Chemically treated wood — pressure-treated lumber, painted wood, manufactured board products — should never be composted. Treated wood often contains copper chromium arsenate (CCA) or other preservatives that persist in compost and accumulate in soil. Glossy paper and cardboard with synthetic coatings resist decomposition and may contain heavy metals.

Plastics and metals do not compost. This is self-evident, but worth stating because a significant source of microplastic contamination in home compost is stickers on fruit and vegetables. Remove produce stickers before adding to the pile — they are typically made of polypropylene and do not decompose.

Quick Reference: What to Add and What to Avoid

Add freely: vegetable and fruit scraps, coffee grounds, tea leaves, eggshells, grass clippings (in thin layers), garden trimmings, dried leaves, shredded newspaper, cardboard, wood chips, straw, hay, aged manures, seaweed (rinsed), natural fiber fabrics, hair, feathers.

Add with care: fresh manures (hot-compost to pathogen-kill temperature), citrus in large quantities (slows decomposition slightly), onion and garlic in large quantities (can discourage worms), sawdust in large quantities (very high C:N — balance with nitrogen inputs), wood ash (use sparingly).

Do not add: meat, fish, dairy, cooked food with oils, pet waste from carnivores, human waste in standard systems, chemically treated or painted wood, glossy paper, plastics, metals, diseased plant material unless confident of hot-composting temperatures.

Chapter 8
Building, Maintaining, and Troubleshooting Your Pile

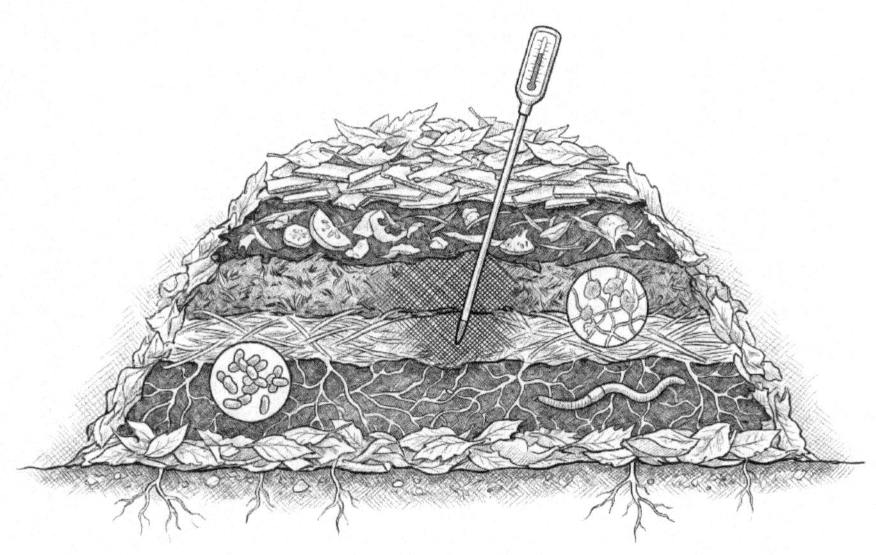

My first solo compost pile was a disaster. I threw in everything — grass clippings, melon rinds, an entire newspaper, some leftover pasta — and let it sit for six months. What I came back to was a matted, half-decomposed mass that smelled like a wet dog. The pasta had attracted rats. I had to start over. That failure taught me more about composting than the first book I read on the subject.

What I learned was not a longer list of rules. It was a set of principles that made the rules make sense. Once you understand why each step matters, troubleshooting becomes straightforward: you observe the pile, identify what it's telling you, and address the specific cause rather than guessing.

Building the First Pile

Start with a base layer of browns — coarse material like wood chips, straw, or small twigs — four to six inches deep. This layer serves two functions: it provides carbon to balance the first addition of greens, and it elevates the pile above the ground, improving drainage and allowing air movement from below.

Add your first layer of greens — kitchen scraps, fresh grass clippings, garden trimmings — two to three inches. Then another layer of browns. Continue layering, maintaining roughly a two-to-one ratio of browns to greens by volume. This approximates a 25 to 30:1 C:N ratio for most common materials.

Moisten as you build. Each layer should feel like a wrung-out sponge — damp but not dripping. Dry materials stacked without moisture will not support the microbial activity you need. If your materials are already moist — like fresh kitchen scraps or green grass clippings — reduce or eliminate additional water in those layers.

Size determines performance. A pile smaller than roughly three feet on each side won't retain enough heat to go thermophilic. A pile significantly larger than five feet on each side becomes difficult to turn and may develop anaerobic zones in the center. The sweet spot for most home composters is three to four feet wide and three to four feet tall, with depth matching width.

Ongoing Maintenance

Once built, a pile needs moisture monitoring and turning. The frequency of both depends on how actively you want to manage it.

In hot composting, temperature drives the turning schedule. Check temperature every two to three days for the first two weeks. When the thermometer reads at 131°F or above, the pile is in its thermophilic phase. When temperature begins to drop — typically after five to seven days — turn the pile to reintroduce oxygen and move outer material to the center. After two or three complete temperature cycles, the active phase is complete and the pile enters curing.

In casual composting — the approach most household composters actually use — turn the pile when you think of it. Every two to four weeks is adequate for reasonable decomposition speed. Each turning speeds the process; the absence of turning simply slows it.

When adding new materials to an existing pile, bury them in the center rather than placing them on top. This reduces pest attraction, speeds their incorporation into the pile's microbial community, and maintains better temperature in the active zone.

Seasonal adjustments are covered in depth in the next chapter, but the short version: summer requires more attention to moisture, winter requires more attention to insulation, and spring and fall bring their own specific management priorities.

Troubleshooting: The Five Most Common Problems

Slow decomposition has four common causes. The pile may be too dry — check moisture and add water if needed. It may be too cold for the season, in which case insulating it or waiting for warmer weather is the answer. The C:N ratio may be off — if the pile is very brown and dry-looking, add nitrogen in the form of fresh greens or diluted liquid fertilizer. Or the pile may simply be too small to retain heat. Determine which cause applies before applying a solution.

Unpleasant odor requires diagnosis by smell type. An ammonia smell — sharp, eye-watering, nose-prickling — means too many greens relative to browns. The excess nitrogen is volatilizing as ammonia gas. Add browns and turn the pile to release the excess. A rotten or sulfur smell — the smell of a drain, or a wet garbage bin — indicates anaerobic conditions. The pile is too wet or too compacted, and decomposition has gone anaerobic. Turn it thoroughly, add dry browns, and ensure drainage is adequate.

Pest activity depends on which pest. Rodents are attracted to food scraps, especially protein-rich items. Ensure no meat, dairy, or cooked food is in the pile. Burying kitchen scraps in the center of the pile rather than leaving them on the surface reduces attractiveness. A wire mesh barrier at the base of the pile prevents burrowing. Fruit flies are attracted to exposed fruit and vegetable scraps. Burying inputs and maintaining a carbon-rich surface layer eliminates most fruit fly problems. Ants indicate a dry pile — restore moisture and they will relocate.

A pile that is too wet and has begun to mat compactly has lost its pore structure and is likely partially anaerobic. Turn it vigorously, adding dry browns as you go. If the pile is very heavy and wet, spread it out temporarily to dry before reassembling. Drainage from the base is the long-term solution.

A pile that never heats up may be too small, too dry, too high in carbon, or lacking nitrogen sources. Work through the diagnostic list in order: check volume, check moisture, check the visual ratio of browns to greens. If everything seems right, add a nitrogen activator — fresh grass clippings, chicken manure, or a diluted nitrogen fertilizer — and turn the pile.

Tools That Make a Difference

A compost thermometer is a long-stemmed probe thermometer, typically 16 to 20 inches long, with a dial or digital readout at the handle. A student at one of my workshops once asked if a compost thermometer was strictly necessary. I said no, and then I said: but once you've used one and watched a pile climb from ambient temperature to 140°F over three days, you'll never not want to know what's happening in there. It's a bit like learning to read a recipe versus just guessing. Both produce food. Only one produces consistently good food.

A pitchfork is the essential turning tool. A four-tined pitchfork moves compost effectively without compacting it the way a shovel would. Handle length matters — look for a handle that allows you to work without bending at the waist, which puts the effort into your legs and shoulders rather than your lower back.

A compost aerator — a pole with collapsible wings at the tip that open as you pull it out — is useful for adding oxygen to a pile without a full turning. Push it into the pile, pull it out, and the wings pull air into the channel left behind. They are helpful for piles that are too wet to turn easily, or for a quick aeration between full turnings.

A kitchen pail or countertop collector is the container that lives on your kitchen counter or under the sink, collecting scraps for two to three days before a trip to the main pile. Look for a model with a tight-fitting lid and a charcoal filter to manage odors. The convenience of collecting scraps near the kitchen significantly increases how consistently you add to the pile.

Chapter 9
Seasonal Composting: Working With Nature's Rhythms

Cold composting is slow — no question about that — but slow and reliable beats fast and abandoned every time. The composter who manages a passive pile through all four seasons, making small seasonal adjustments and harvesting once a year, produces more compost over a decade than the enthusiast who attempts hot composting in July, abandons it in August, and doesn't restart until the following spring.

The seasons don't pause your composting operation — they change its character. Understanding what happens in your pile during each season lets you work with those changes rather than against them.

Spring

A compost pile that went dormant or semi-dormant over winter is not dead. The microbial populations are there, waiting for temperatures and moisture conditions that allow them to resume activity. The first warm days of spring — particularly when combined with the moisture from snowmelt or spring rain — reactivate a winter pile rapidly.

The first spring task is a thorough turning. This does several things: it exposes the interior to warming spring air, redistributes materials that may have become unevenly frozen or wetted, and introduces fresh oxygen to kick-start aerobic activity. Add a layer of fresh greens as you turn — the first grass clippings of the season, some kitchen scraps accumulated over winter — to provide nitrogen and restart microbial activity.

Spring also brings the first surge of fresh green waste from the garden: early prunings, cleaned-up perennial growth, the first grass cuts. This material is nitrogen-rich and requires balancing with browns. If your winter pile is high in carbon-heavy material (dried leaves, cardboard), spring greens will help balance it. If you're adding primarily high-nitrogen spring greens, keep dry browns on hand.

Overwintered compost that has cured through the cold months is often ready to apply in spring, before planting. This is ideal timing — applying compost before planting incorporates it into the soil structure that seedling roots will encounter first.

Summer

Summer creates two management challenges: drying and excess nitrogen. High temperatures and low rainfall dry a pile faster than it can absorb moisture from new inputs. An outdoor pile in July can go from well-moistened to dangerously dry in two weeks without supplemental water.

Check moisture weekly in summer, more often during heat waves. If the interior feels warm but dry when you probe it with a gloved hand, water it immediately and turn it to distribute moisture throughout. A layer of straw or dried leaves on top of the pile acts as mulch, reducing surface evaporation.

Summer gardens also produce high-nitrogen waste in large quantities: grass clippings, annual flower trimmings, squash leaves, tomato suckers. A pile receiving large quantities of fresh green material in summer needs proportionally more brown material — and enough turning to prevent matting.

The good news: summer heat accelerates decomposition. A well-managed summer pile can move from freshly built to finished compost in six to eight weeks. Monitor temperature, maintain moisture, and turn when needed.

Autumn

Autumn is composting's most productive season. The abundance of fallen leaves provides carbon material in quantities that most summer piles lack. If you collect and store autumn leaves, you have a brown reserve that will balance your pile's nitrogen inputs throughout the following year.

Managing leaf volume is the primary autumn challenge. A single large tree can produce more leaves than a home compost pile can absorb without becoming overly carbon-heavy. The solution is to run a parallel leaf-mold operation: a separate pile or bin where leaves are wetted and left undisturbed to break down slowly over one to two years. Leaf mold doesn't provide nutrients, but it improves soil structure significantly and is an excellent mulch material.

Autumn is also the time to build up the active pile's mass before winter. A larger pile loses proportionally less heat to the environment — the surface-to-volume ratio decreases as size increases. Building the pile to its maximum size before the first frost gives it the best chance of staying active, or at least of reaching the hottest internal temperature before winter suppresses activity.

In autumn, prepare the pile for winter: ensure it's adequately moist before the ground freezes (dry material is harder to rehydrate in winter), check that its structure allows air movement without being so open that it loses heat rapidly.

The Transition Periods

The calendar divisions between seasons are convenient fictions. The pile doesn't receive a signal on the first day of autumn that it should begin preparing for winter. What it receives instead is a gradual shift in temperature, moisture, and available inputs — and the composter who pays attention to those signals, rather than the date, manages a more responsive pile.

Late summer into autumn is the most consequential transition of the composting year. Daytime temperatures are still warm enough to sustain active decomposition, but the character of available material is shifting: high-nitrogen summer greens are thinning out, the first leaves are beginning to fall, and the garden is moving toward harvest rather than growth. This is the window to build the pile's mass deliberately — to take advantage of the residual warmth while incorporating the first carbon-rich autumn material. A pile built up in September, when the soil is still warm and biological activity is still high, will have established a strong microbial population before the first frost arrives.

The late winter into early spring transition is subtler but equally important. A pile emerging from a cold winter thaws unevenly — the surface may still be frozen while the interior has already begun to warm. Probe the pile with a compost thermometer before assuming it's still dormant. If the interior has risen above 40°F, microbial activity has already resumed. Adding the first greens of the season at this point — rather than waiting for spring to feel fully established — gives those reactivating microbes the nitrogen boost they need to accelerate activity quickly.

Winter

What actually happens in a frozen pile? The short answer: the microbial community goes dormant, but doesn't die. Psychrophilic bacteria continue slow activity even below freezing. The physical structure of the pile is preserved. When temperatures rise, activity resumes. A frozen pile is not a failed pile — it's a pile in suspension.

In mild winters — average temperatures above 40°F — an insulated pile can maintain some microbial activity. Straw bales placed around the pile on three sides, or a layer of straw or burlap on top, significantly reduce heat loss. Some composters use a black polyethylene tarp on top, which absorbs solar radiation on clear winter days and helps maintain interior temperature.

In cold climates with extended freezes, the most practical approach is to collect inputs over winter and add them to the pile in spring. A covered storage bin for kitchen scraps — or simply freezing them — allows you to accumulate material through winter without adding it to a frozen pile where it will sit unprocessed until spring.

For year-round composting in cold climates, indoor systems are the reliable answer. A worm bin in a warm basement operates regardless of outdoor temperature. A bokashi bin on the kitchen counter processes food waste through fermentation with no temperature sensitivity. These systems are covered in detail in Books III and IV.

The Composting Year and the Garden Calendar

Composting and gardening share the same seasonal rhythm, and aligning them deliberately produces better results in both.

In spring, the priority use for any compost that overwintered and cured through the cold months is soil preparation before planting. Compost incorporated into beds in March or early April — before seeds go in or transplants go out — gives soil organisms time to begin integrating the organic matter into the soil structure. The seedling roots that push into that amended soil in May encounter a more biologically active, better-structured growing medium than roots pushing into unamended soil.

In summer, finished compost serves a different function: side-dressing. Heavy-feeding crops — tomatoes, corn, squash, brassicas — benefit from compost applied in a ring around the plant's drip line midway through the growing season. This is not soil amendment in the structural sense; it is a feeding, and it should be timed to coincide with the period of maximum growth and fruit development.

In autumn, post-harvest incorporation is the most valuable application of the season. After annual crops are cleared, working compost into the top six to eight inches of soil allows it to break down slowly over winter and integrate fully before the following spring planting. Autumn-applied compost that sits under a cover crop or a layer of mulch through winter is typically better integrated into the soil food web by spring than compost applied fresh in March.

In winter, the composting calendar and the garden calendar share a task: planning. A winter review of what the garden produced, what the pile processed, and what inputs are available for the coming season allows adjustments before they are urgently needed. A gardener who enters spring knowing they have a nitrogen deficit in their pile — because the previous year was heavy on leaves and light on kitchen scraps — can compensate from the first week of the season rather than discovering the problem in June.

Chapter 10

Knowing When It's Ready: Harvesting and First Use

S omeone once asked me what finished compost should smell like. I said: a forest floor after the first rain of October. They nodded as if they knew exactly what I meant. A few people will. For the rest, it's one of those things where you'll know it when you smell it — and once you do, you'll never confuse it with anything else.

Knowing when compost is ready is more important than knowing when to harvest. Compost applied before it's ready can set your plants back rather than helping them. Patience here is not passive — it is the step that makes everything else worth doing.

Signs of Maturity

Visually, finished compost is dark brown to black, crumbly in texture, and homogeneous. No original materials are recognizable. There are no distinct layers, no intact vegetable peels, no leaf fragments that retain their original shape. The material has a uniform granular quality that resembles the best garden soil.

Olfactorily, the smell is earthy and forest-floor-like. There is no sourness, no ammonia, no sulfur, no hint of rot. If any of those smells are present, the pile needs more time — and in the case of ammonia or sulfur, probably some turning and adjustment.

Tactilely, finished compost is spongy and slightly moist, not slimy, not powdery. It holds together when you squeeze it but crumbles when you open your hand. Slimy compost has been too wet during curing. Powdery compost has dried out.

The bag test is the most reliable field method. Take a cup of compost from the center of the pile, seal it in a plastic bag, and leave it at room temperature for three days. Open it. If it smells earthy, the compost has cured completely. If it smells sour, ammonia, or off in any way, return it to the pile for more curing time.

How Much to Expect

A common question from first-time composters is how much finished compost a pile actually produces. The answer is less than most people expect: organic material loses the majority of its volume during decomposition. A cubic meter of mixed greens and browns will typically yield 200 to 300 liters of finished compost — roughly one quarter to one third of the original volume. The reduction is greater with nitrogen-rich inputs, which break down more completely, and smaller with woody or fibrous materials that retain more structure. Plan accordingly: a single pile rarely produces enough to amend an entire garden. It produces enough to do the most important parts of it well.

Dealing with Uneven Maturity

Most piles don't finish evenly. The center may be beautifully cured while the outer edges still contain recognizable materials. This is normal and doesn't require waiting for the entire pile to reach the same state.

Screening separates finished from unfinished compost reliably. Build or buy a simple screen — hardware cloth with half-inch openings mounted on a frame — and sieve the pile through it. Finished compost passes through. Coarse, unfinished material stays on the screen and goes back into the next pile. This approach lets you harvest partially cured batches without waiting for complete uniformity.

Screened compost is also a more uniform product for fine applications — seedling potting mix, top dressing lawns — where the granular texture matters. Unscreened compost is fine for mulching or broad soil incorporation.

Using Almost-Ready Compost

Material that doesn't pass the bag test isn't wasted — it's compost that needs a different application. Partially decomposed material can be used safely as a surface mulch around established trees and shrubs, where it will continue breaking down in place without contacting root zones directly. It can also serve as a base layer in a new lasagna bed, where further decomposition is part of the design. What it should not do is go directly into planting holes, seedling trays, or any application where immature organic matter will be in close contact with active root systems. Nitrogen theft and phytotoxic compounds from incomplete decomposition are real risks — but only when the material is placed where roots will encounter it immediately.

Harvesting Methods

Harvesting from a stationary bin depends on the bin's design. Most purchased plastic bins have a small door or hatch at the base. The compost harvested from here is the most mature, having settled to the bottom over time. This is a practical but slow method — access is limited and output volume per session is small.

For a DIY bin, the most efficient harvest method is to disassemble one side of the bin, exposing the pile. Remove outer material that hasn't finished composting into a temporary pile or directly into a new bin. Harvest the dark, crumbly interior. Reassemble the bin around the unfinished outer material and continue.

For an open pile, the process is similar: push the outer material aside, harvest the finished core, and rebuild with the unfinished outer material as the new pile's base.

Tumblers typically produce batch compost: the entire drum contents are ready at approximately the same time. Stop adding new material to the drum when it's full, continue turning for the curing period (two to four weeks after active decomposition slows), then harvest the entire batch. Start the next batch fresh. Some tumblers have two independent chambers to maintain continuous production.

Storing Finished Compost

Finished compost doesn't need to be used immediately, but it does need to be stored correctly. Left exposed to heavy rain, compost loses soluble nutrients through leaching. Left to dry completely in direct sun, it loses microbial activity. The ideal storage condition is cool, slightly moist, and covered — a lidded bin, a pile under a tarp, or a covered corner of a shed. Stored this way, finished compost holds its quality for six to twelve months without significant loss. Beyond that, it remains useful as a soil conditioner but has lost much of its microbial richness. Use it within a season when possible.

First Use: Getting the Most from Your Initial Batch

Top dressing is the simplest application: spread one to two inches of finished compost over the surface of established garden beds and let rain and soil organisms incorporate it. This is appropriate for established perennial beds, around shrubs and trees, and on lawns in spring and fall.

Mixing into planting holes is the highest-value application per unit of compost. When transplanting seedlings or planting bare-root plants, mix a generous quantity of compost into the planting hole — up to 20 to 30% by volume of the backfill mix. This puts fertility exactly where new roots will encounter it first.

A simple compost tea for seedlings can be made by mixing a cup of finished compost in a gallon of water, stirring, and letting it steep for a few hours before watering seedlings with the filtered liquid. That's not aerated compost tea — it's a basic extract — but it delivers soluble nutrients and some microbial inoculation to new transplants. Proper aerated compost tea is covered in Book V.

For your first harvest, prioritize your most valuable planting areas: the vegetable bed, the planting holes for new trees or shrubs, the seedling trays if you grow your own transplants. A small first harvest used precisely will do more good than the same quantity spread thin across areas that don't need it.

Chapter 11

Safety First: Pathogens, Ergonomics, and Practical Precautions

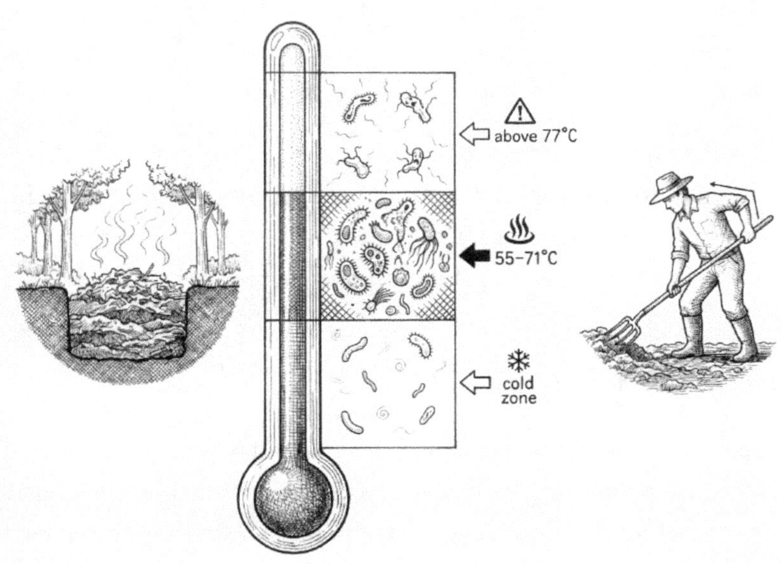

M ost composting is completely safe. That needs to be said first, clearly and without qualification. The vast majority of home composters manage their piles with no adverse health effects over years and decades of regular practice. What follows is about knowing which risks are worth managing and which are effectively zero — so you can work without unnecessary caution and without unnecessary carelessness.

Two categories of risk are relevant to home composting: biological, in the form of pathogens that can survive inadequate processing, and physical, in the form of ergonomic strain from the manual work involved. Both are manageable.

Understanding Pathogens in the Compost Pile

Pathogens enter compost piles through the input materials. The most significant route is raw or undercomposted animal manure, particularly from non-ruminant animals. E. coli O157:H7 and various Salmonella species are the pathogens of greatest concern in composting contexts. Both are heat-sensitive.

E. coli O157:H7 is destroyed at 131°F (55°C) within minutes. Salmonella spp. require the same temperature range. The USDA standard for pathogen destruction in composted materials requires that the pile reach 131°F (55°C) and maintain that temperature for a minimum of three consecutive days. This is achievable in a well-managed hot pile and provides a reliable safety margin.

Thermophilic composting at the temperatures described in Book I Chapter 3 destroys these pathogens reliably. Cold composting, which does not reach thermophilic temperatures, cannot guarantee pathogen destruction and should not be used for manure inputs or disease plant material destined for food garden applications.

Managing High-Risk Materials

Fresh manure: all animal manures should be composted before application to food garden beds. The composting process — specifically the thermophilic phase — destroys the pathogens that fresh manure may contain. Aged manure that has been stored in a pile for several months without active hot composting is not the same as thermophilically composted manure. If you are not certain your manure pile has reached pathogen-kill temperatures, treat it as unprocessed.

Pet waste from dogs and cats carries specific human-transmissible pathogens. Standard home composting is not an appropriate processing method for this material. Dedicated pet waste composters — separate from the main pile, not used for food garden soil — exist for households that want to manage this stream responsibly.

Meat and dairy: if you add these to your compost pile through inadvertent contamination — a piece of food you didn't notice — it's not a crisis. The issue with meat and dairy in a compost pile is not primarily pathogen risk in a thermophilic pile; it's pest attraction. A pile containing meat scraps that doesn't reach thermophilic temperatures quickly is an invitation to rodents. Bokashi fermentation, covered in Book III, is the appropriate method for these inputs.

Safe Use of Finished Compost

Finished compost that has undergone proper thermophilic processing presents minimal pathogen risk. The safety precautions around applying it are sensible but not onerous.

Wash all produce that has contacted compost or compost-amended soil before eating. This is standard practice regardless of how the soil was managed and applies equally to compost-grown vegetables and conventionally grown ones.

Avoid direct application of compost to the edible portions of crops that will be eaten raw without cooking. Don't spread compost over the top of leafy greens or directly onto root vegetable surfaces. Side-dress rather than top-dress for these crops.

The standard guidance for food crop safety is a 90 to 120-day interval between application of uncomposted organic matter and harvest of food crops with soil contact. Properly composted material does not carry this restriction, but applying compost before planting rather than during the growing season is a simple practice that eliminates any concern.

Physical Safety and Ergonomics

Turning a compost pile is physically demanding work. A cubic yard of wet compost weighs several hundred pounds. Working through it with a pitchfork engages the back, shoulders, and legs. Done incorrectly, it causes exactly the lower-back strain that keeps people away from physical outdoor tasks.

The correct technique is the same as any heavy lifting: bend at the hips and knees, keep the load close to your body, use your leg muscles to drive the movement rather than your back. With a pitchfork specifically, pivot with your whole body rather than twisting at the waist. Take breaks. There's no deadline on turning a compost pile.

Long-handled tools reduce strain by allowing you to work without excessive bending. A pitchfork handle should reach comfortably from the ground to your armpit when upright. If you're consistently bending at the waist to reach the pile, your tool handle is too short for your height.

Dust and mold spores are present in any active compost pile. For most people, exposure during turning is minimal and without consequence. For people with respiratory sensitivities — asthma, mold allergies — wearing a simple N95 mask during turning is a reasonable precaution.

Gloves are worth wearing, particularly when working with fresh manure or recently added kitchen scraps. Standard garden gloves are adequate. Disposable nitrile gloves offer better protection when handling materials of uncertain origin.

In hot weather, treat pile turning as outdoor physical labor: start early in the morning before temperatures peak, stay hydrated, and stop if you feel overheated. It's compost. It will wait.

PART III

MASTERING COMPOSTING TECHNIQUES

A Method for Every Gardener and Every Situation

Chapter 12

Hot Composting: Speed, Heat, and Precision

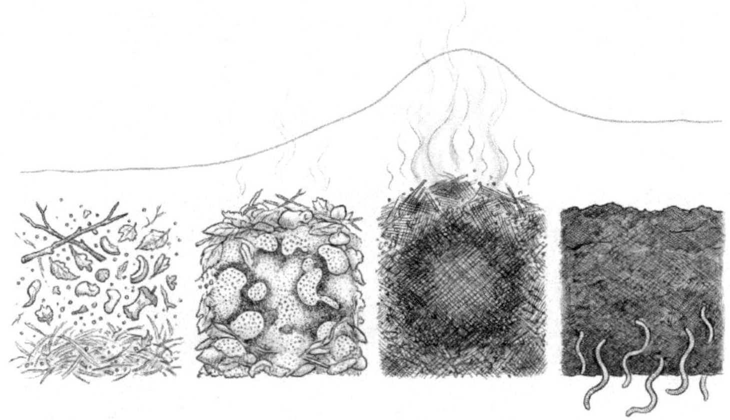

Build the pile on a Tuesday. Check the temperature on Thursday. If you have done it right, the thermometer will read somewhere between 130°F and 150°F — and you will understand, without needing further explanation, why people who compost this way rarely go back to anything slower. Hot composting is the most labor-intensive method in this book. It is also the fastest, the most controllable, and the one that produces the highest-quality output when managed well.

Every other method in Book III is, in some sense, a variation on the conditions that hot composting optimizes. Understanding it in full makes everything else easier to evaluate.

How It Works

Hot composting works by concentrating the conditions that thermophilic bacteria — those operating between 131°F and 160°F (55°C to 71°C) — need to dominate the pile. Those conditions are: sufficient mass to retain heat, a balanced C:N ratio to sustain rapid microbial activity, adequate moisture, and enough oxygen to keep the process aerobic. Get all four right, and the pile heats itself.

The key species belong primarily to the genera Thermus and Bacillus. At their peak, these bacteria reproduce and metabolize so rapidly that a well-built pile can reach 140°F within 24 to 48 hours of construction. That heat is the pile's most visible indicator of health — and its most useful diagnostic tool.

Mass is the factor most commonly underestimated. A pile smaller than one cubic meter (roughly 3 feet on each side) loses heat through its surface faster than thermophilic bacteria can generate it. The relationship is geometric: as pile size increases, volume grows faster than surface area. A cubic meter is the minimum; a pile twice that size heats more reliably and maintains temperature longer through turning cycles.

Building the Pile

The starting point is a 1:1 mix of greens and browns by volume — not by weight. By volume, this approximates a C:N ratio of 25 to 30:1 for most common material combinations. If you have specific C:N data for your inputs (see Book I Chapter 3), calculate the ratio directly and adjust accordingly. The 1:1 volume rule is a reliable starting approximation, not a fixed formula.

Begin with a four-to-six-inch base layer of coarse browns — wood chips, straw, or small woody material — to elevate the pile off the ground and allow some air movement from below. Layer greens and browns alternately in three-to-four-inch layers. Moisten each layer as you build: the interior of the pile should reach 50 to 60% moisture throughout, not just on the surface.

If you are building from stored materials rather than fresh inputs, add a shovelful of finished compost or active garden soil to each layer. This inoculates the pile with a diverse microbial community and can cut the time to first peak temperature by one to two days.

Monitoring and Turning

Check temperature daily for the first two weeks, using a long-stemmed compost thermometer inserted to the center of the pile. Take readings from at least two or three locations — temperature varies within the pile, and a single reading from one spot can mislead.

The target range for pathogen destruction and optimal decomposition is 131°F to 160°F (55°C to 71°C). When the temperature drops below 131°F after an initial peak — typically within five to seven days — turn the pile. Turning does three things simultaneously: it reintroduces oxygen to depleted interior zones, moves cooler outer material to the center where the heat is, and breaks up any matted or partially anaerobic sections.

After turning, a well-built pile will reheat within 24 to 48 hours. When it peaks and drops again, turn it again. Two to three full temperature cycles usually complete the active phase. After the pile fails to reheat above 110°F following a turn, the thermophilic phase is effectively over — shift to curing.

During the active phase, check moisture at every turning. The pile loses water rapidly through evaporation at high temperatures. If the interior feels dry or crumbly, add water as you turn, ensuring it penetrates throughout. A dry pile in the active phase is the most common reason a well-built hot pile fails to reheat after its first cycle.

Curing

Curing is not optional. Finished active-phase compost is biologically active but chemically unstable — intermediate decomposition products are still present, and the pile needs two to four weeks of undisturbed rest to complete the process. During curing, mesophilic bacteria and fungi finish breaking down remaining compounds, pH stabilizes toward neutral, and the physical structure of the compost develops.

Cover the curing pile with a tarp or burlap if rain is likely — you want to maintain moisture without waterlogging. Do not turn it. When it smells like a forest floor and maintains a temperature at or just above ambient for a full week, it is ready for the bag test described in Book II Chapter 5.

Timeline, Strengths, and When to Use It

Under ideal conditions — correct C:N ratio, adequate moisture, sufficient mass, turned every three to five days — hot composting produces finished compost in four to eight weeks. In practice, first-time hot composters often take ten to twelve weeks as they learn to read and adjust the pile. Subsequent batches go faster.

The strengths are specific. Hot composting kills weed seeds and human pathogens reliably when temperatures stay in the target range for the required duration. It produces the highest-nitrogen, most biologically active compost of any method. It converts large volumes of material quickly. For gardeners with substantial yard waste or a pressing need for finished compost by a specific date, nothing else performs comparably.

The demands are also specific. Hot composting requires enough material to build a full pile in one session — you cannot add small amounts daily and expect thermophilic temperatures. It requires monitoring and turning on a schedule that does not accommodate frequent travel or weeks-long absences. And it requires a minimum space that not every garden provides.

Choose hot composting if you have large seasonal volumes of material, need finished compost within two months, want to compost manure or potentially diseased plant material, or simply want the most complete control over the process. If your waste stream is small or intermittent, cold composting or vermicomposting will serve you better.

Chapter 13

Cold Composting: The Patient, Low-Effort Approach

There is a version of composting that requires almost no management, produces no heat worth measuring, never needs turning on a schedule, and still converts organic matter into something your garden can use. The trade is time: months instead of weeks, sometimes a year or more instead of four to eight. For a significant portion of composters — the busy, the space-constrained, the disinclined toward outdoor schedules — that trade is straightforwardly worth making.

Cold composting is not a degraded version of hot composting. It is a different process, managed by different organisms, with different strengths and a distinct appropriate use. Understanding it on its own terms, rather than as hot composting without the effort, is what lets you deploy it well.

How It Works

Where hot composting is dominated by thermophilic bacteria operating at high metabolic rates, cold composting is primarily a mesophilic process — temperatures stay at or near ambient, between 50°F and 110°F (10°C to 43°C), and the organisms doing the work are a broader and slower-moving community. Bacteria are present, but fungi play a proportionally larger role than in a hot pile. Larger organisms — earthworms, beetles, millipedes, mites — contribute more to the physical breakdown of materials.

Because temperatures never reach thermophilic levels, cold composting does not kill weed seeds or pathogens. This is the most practically significant limitation. Weed seeds added to a cold pile will germinate when the compost is applied to the garden. Diseased plant material should not go into a cold pile unless you are certain it will be used only on established beds, not in seed-starting mixes or planting holes.

Everything else decomposes. The process is simply slower, and the output — while nutritionally sound — tends to be less biologically active than the output of a hot pile. The microbial population is there; it has just not been through the concentrated thermophilic enrichment that hot composting produces.

Setting Up and Managing a Cold Pile

The setup is minimal. Any container or open heap works. There is no minimum size requirement in the way there is for hot composting — a cold pile will decompose regardless of mass, just at a rate proportional to its volume and the ambient conditions. A simple wooden bin, a cylinder of wire mesh, or a designated corner of the garden all work equally well.

Add materials as they accumulate. There is no layering formula to follow, no schedule to maintain. Kitchen scraps, garden trimmings, leaves, cardboard — add what you have when you have it. The pile will process everything eventually.

The one management step that makes a measurable difference is occasional turning — even infrequent turning, once every month or two. Each turn introduces oxygen to the interior, accelerates decomposition in the outer zones that have not yet been processed, and redistributes moisture. A cold pile turned four or five times per year produces finished compost noticeably faster than one left entirely undisturbed, without becoming the management commitment that hot composting requires.

Moisture management matters even in a cold pile. A pile that dries out stops decomposing. One that becomes waterlogged goes anaerobic. The target range — 50 to 60% moisture — applies here as much as in hot composting; the pile simply tells you more slowly when it's wrong.

Bottom Harvesting

In a cold pile where new material is added continuously at the top, the most mature compost is always at the bottom. The pile stratifies naturally by age: fresh material on top, progressively older and more decomposed material working downward. This creates a practical harvesting approach without waiting for the entire pile to finish.

To harvest from the bottom of an open heap, tip or disassemble the pile, set aside the upper material that is still recognizable, and collect the dark, crumbly bottom layer. Rebuild the pile with the upper material and continue. In a bin with a bottom access door, harvest through the door a few shovelful at a time as the lower material matures, without disturbing the active upper zone.

The compost from the bottom of a cold pile is generally ready within six months to a year, depending on materials, climate, and how often the pile is turned. Leaf-dominant piles take longer than kitchen-scraps-dominant ones. Cold climates extend timelines by several months.

Timeline, Strengths, and When to Use It

Six months to two years is the honest range, with most well-maintained cold piles producing harvestable material in nine to twelve months. The variability is large because the factors that determine speed — material type, moisture, turning frequency, temperature — are all more variable in a passively managed system.

Cold composting handles small, intermittent waste volumes that would never accumulate to the minimum mass needed for a hot pile. It suits households that generate kitchen scraps consistently but lack the yard waste volume for periodic large builds. It requires no schedule and tolerates periods of neglect that would stall a hot pile. It is beginner-friendly precisely because there is no failure mode beyond extreme dryness or flooding: the pile will process materials given enough time regardless of how imprecisely it is managed.

The constraints are the ones stated clearly at the outset: slow, no pathogen kill, lower biological activity in the output. Cold composting works alongside other methods more often than it replaces them. A cold pile running continuously in one corner of the garden, processing leaf litter and minor kitchen waste, complements a hot pile that handles larger seasonal volumes and manure inputs.

Chapter 14

Vermicomposting: Harnessing the Power of Worms

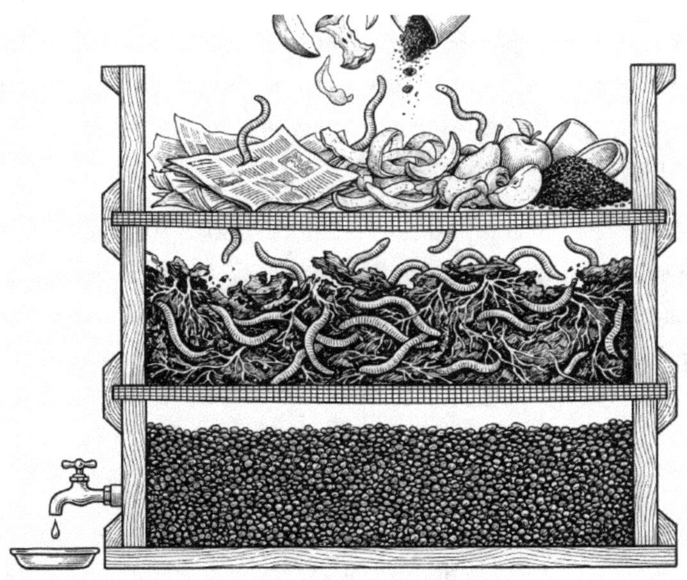

Red wigglers eat half their body weight in food waste every day. A pound of worms — roughly 800 to 1000 individuals — processes half a pound of kitchen scraps in that same period. They do this continuously, producing castings that are more nutritionally concentrated than standard compost, richer in beneficial microorganisms, and immediately available to plants. A well-managed worm bin is among the most productive composting systems per square foot that exists. The footprint can be smaller than a kitchen cabinet.

Vermicomposting is the method that surprises people most when they actually try it. It looks low-tech — a box of worms eating your banana peels — and then the castings come out and the seedlings you pot in them grow noticeably differently from the ones in plain potting mix.

The Biology

Worms are not, strictly speaking, the primary decomposers in a worm bin. They are the processing layer on top of a dense microbial community. Worms grind food through a muscular gizzard, extracting nutrition, and excrete castings packed with bacteria, fungi, and other organisms at concentrations far higher than in surrounding material. Much of what worms eat is not food directly — it is the microbial film coating the food. The organisms in the gut and the castings are what make vermicompost nutritionally exceptional.

A cubic centimeter of vermicompost can contain 10,000 to 100,000 bacteria — an order of magnitude more than typical finished compost. The diversity is also higher. Worm bins that receive varied inputs develop microbial communities that reflect that variety, and the castings inoculate garden soil with a broader range of organisms than most other amendments deliver.

Choosing Your Worm Species

Eisenia fetida — the red wiggler, also called redworm or brandling worm — is the standard choice for home vermicomposting everywhere in the temperate world. It thrives in the exact conditions a worm bin creates: confined space, organic-rich substrate, temperatures between 55°F and 77°F (13°C to 25°C). It does not burrow deep into soil and will not escape into your garden if the castings are applied outdoors. It breeds readily in captivity.

For cellars, garages, or any unheated space that stays cool through winter, Eisenia hortensis — the European nightcrawler — handles the lower end of the temperature range better than red wigglers do. It processes coarser material and produces larger castings, though more slowly. If fishing bait is a secondary use, it is also the better choice on that count.

Two tropical species — Eudrilus eugeniae (the African nightcrawler) and Perionyx excavatus (the Indian blue) — both outperform red wigglers in raw throughput, but only reliably in warm climates. In temperate conditions they underperform. The Indian blue has an additional drawback worth knowing about: it responds to even minor environmental stress with coordinated escape attempts, which makes it considerably less forgiving to learn on.

Setting Up the Bin

Bin size is determined by your weekly food waste volume. The formula is one square foot of surface area per pound of weekly food waste. A household generating three to four pounds of kitchen scraps per week needs a bin with three to four square feet of surface area — roughly the footprint of a standard under-sink cabinet. Depth matters less than surface area: worms are surface feeders and will not populate the lower third of a deep bin.

Drill aeration holes in the lid and upper sides — quarter-inch holes, spaced two to three inches apart. The base needs drainage holes as well; set the finished bin on a tray to catch leachate, because without that outlet, waterlogging drives worms to the surface and eventually kills populations.

Bedding is the material that fills the bin before worms arrive and constitutes the ongoing substrate they live in. Shredded newspaper or cardboard, dampened to roughly 80% moisture (wet but not dripping), is the standard. Coconut coir works equally well and holds moisture more consistently. Fill the bin two-thirds full with damp bedding before adding worms. The bedding is not waste — it is the worms' immediate habitat and will itself eventually become part of the castings.

When introducing worms, spread them across the surface of the bedding in dim light. Red wigglers have a strong light-avoidance response and will move downward into the bedding within minutes. Leave the lid off for an hour or two to encourage settlement. Do not feed for the first three to four days — let the worms acclimate to their new environment before introducing additional stress.

Feeding and Ongoing Management

Feed by burying scraps in a different section of the bin each time, rotating through quadrants. Burying prevents fruit flies from accessing the surface, reduces odor, and draws worms to a concentrated food source rather than distributing uneaten material across the bin. Push bedding aside, deposit scraps, and cover.

Worms eat most plant-based kitchen waste: vegetable and fruit scraps, coffee grounds, tea leaves, crushed eggshells, bread, cooked grains. Avoid meat, fish, dairy, and oily foods in a standard bin — these attract pests and create anaerobic conditions. Citrus and onion in small quantities are tolerated; in large quantities they raise acidity to levels worms find uncomfortable.

The bin should smell like garden soil — earthy, faintly sweet. A foul smell indicates something is wrong: usually overfeeding (more food than the worm population can process, leading to anaerobic decomposition), excessive moisture, or a pH problem. If the bin smells off, stop feeding for a week, add dry bedding to absorb excess moisture, and let the worm population catch up with the existing food.

Troubleshooting

The first time I set up a worm bin, I woke up the next morning to find several dozen red wigglers attempting to escape across my kitchen floor. I had the moisture level too high and the pH slightly off, and they'd decided en masse that anywhere else was preferable. The lesson: worms are better diagnostic tools than any instrument. When they want out, something is wrong. The question is what.

Mass escape attempts have three common causes: moisture that has become too high or too low, pH that has shifted outside the comfortable range (below 6 or above 8), or introduction of something the worms find toxic — certain pesticide residues, highly acidic foods in quantity, or excessive salt. Identify the cause before returning escapees. If you simply put them back into unchanged conditions, they will leave again.

The fix for fruit flies is straightforward: bury every addition under bedding before closing the lid. They breed on exposed organic matter and cannot reach food buried even an inch below the surface. A piece of moistened cardboard laid flat on the bedding and replaced every week or two eliminates most populations within a fortnight — no sprays, no traps needed.

A mite population usually signals two things happening at once: moisture that has climbed too high and a feeding routine that has been too protein-heavy. Pull back on dense, wet inputs, add dry bedding to bring moisture down, and the mites regulate themselves within a week or two. They are not harmful to worms — just a reliable sign that the balance needs adjusting.

"Protein poisoning" or sour crop is a condition caused by fermenting protein-rich inputs — usually meat or excess grain — producing ammonia and lactic acid that kill worms rapidly. Affected worms flatten and become motionless. Remove the source immediately, add significant amounts of dry brown bedding, and check whether any worms have survived in cleaner zones of the bin. If the condition is caught early, populations recover.

Harvesting Castings

Two methods work reliably. The dump-and-sort method involves emptying the bin onto a plastic sheet in bright light, forming the contents into several cone-shaped piles, and waiting. Worms move away from light, burrowing into the center of each cone. After ten to fifteen minutes, scrape the outer layer of castings away from each cone. Repeat until the piles have shrunk to dense clusters of worms, which go back into the bin along with fresh bedding.

The migration method is less disruptive. Push all existing contents to one side of the bin. Fill the empty side with fresh, damp bedding and begin feeding exclusively on that side. Over four to six weeks, worms migrate to the new food source. Harvest the now mostly-worm-free original side.

Worm tea — properly called leachate when it drains from the bin without aeration — collects in the tray beneath the bin. Dilute it ten-to-one with water before use.

Undiluted leachate can be phytotoxic. Aerated worm tea, made by aerating finished castings in water for 24 to 36 hours, is a different and more biologically potent product covered in Book V Chapter 3.

Timeline and Ideal Use

A new bin with an established worm population produces harvestable castings in two to three months. After that, harvesting is continuous: as worms process material, castings accumulate, and the bin requires regular partial harvesting to maintain adequate working space.

Vermicomposting suits indoor or small-space situations, year-round composting in cold climates, and kitchen-focused waste streams. The castings are particularly valuable for seedling production, potted plants, and targeted soil amendment where the high cost of vermicast from commercial sources makes home production economically worthwhile. A small apartment-scale bin producing a pound of castings per week saves roughly $25 to $40 in purchased castings monthly — at current commercial pricing.

Chapter 15

Bokashi: Fermentation as a Path to Fertility

Pickling preserves food by creating an acid environment hostile to putrefactive bacteria. Sourdough bread, kimchi, miso, sauerkraut — all are products of controlled fermentation, where specific microorganisms acidify their substrate before anything can rot. Bokashi applies the same principle to food waste: instead of decomposing kitchen scraps, it ferments them, stabilizing the material in a preserved state that can then be processed by soil organisms at the second stage.

The reason this matters is that fermentation handles inputs that aerobic composting cannot: meat, fish, dairy, cooked food, citrus in any quantity, heavily spiced leftovers. If your household generates the full range of food waste rather than just the vegetable-centric stream that conventional composting accommodates, bokashi is the system that closes the gap.

What Bokashi Is — and What It Is Not

Bokashi is anaerobic fermentation, not aerobic decomposition. The microorganisms at work — primarily lactic acid bacteria, yeasts, and phototrophic bacteria collectively called Effective Microorganisms (EM) — do not break down organic matter through digestion the way composting bacteria do. They preserve it, in a pickled state, through acidification.

The output of a two-week bokashi cycle is pre-compost: fermented organic material that has been stabilized but not decomposed. It still looks like food. It smells like a slightly sweet fermented product — not unpleasant, but distinctive. To complete its breakdown, it needs contact with soil or a traditional compost pile, where bacteria and fungi finish the process in two to four weeks.

This two-step nature is the most common source of confusion about bokashi. People bury the output, check back in a week, find something that still resembles food, and conclude that the system isn't working. It is working — the fermentation has succeeded. The second step just takes longer than most instructions suggest.

Origins and Science

The modern EM system was developed in the 1980s by Dr. Teruo Higa at the University of the Ryukyus in Okinawa, Japan. Higa isolated and cultured combinations of microorganisms that had been observed improving soil health when naturally present in high concentrations. Traditional East Asian fermentation practices — particularly the long history of Japanese and Korean fermented agricultural amendments — provided the conceptual foundation that Higa's research formalised.

The science of EM is well established for food fermentation but more contested for agricultural applications. What the evidence supports clearly is that bokashi fermentation produces organic acids — lactic acid primarily — that create a hostile environment for putrefactive bacteria, prevent decomposition from going rotten, and result in a stable, nutrient-preserved pre-compost that improves soil structure and biological activity when incorporated. The more expansive claims made by some EM advocates about plant growth promotion and disease suppression are supported by variable evidence.

Equipment and Bokashi Bran

A bokashi bin needs two features: an airtight lid and a spigot or drainage tap at the base. Airtightness is essential — any air entering the bucket disrupts the anaerobic fermentation and allows mold to develop. The drainage tap lets you collect bokashi juice, which accumulates as liquid drains from the fermenting material.

Bokashi bran is wheat or rice bran inoculated with EM and dried. It provides the microbial community that drives fermentation. Commercial bokashi bran is widely available online and at garden centers. Making it yourself requires EM culture (available from suppliers), molasses, bran, and water, mixed and sealed for two weeks before use — a straightforward process if you want to reduce ongoing supply costs.

Step-by-Step Process

Add food waste to the bin in layers of roughly two to three inches. After each addition, sprinkle a layer of bokashi bran — about one tablespoon per pound of waste is the working ratio, though slightly more is better than less in the early learning period. Press the waste down firmly with a plate or flat lid to expel air pockets from within the material. Seal the bin tightly after every addition.

Drain bokashi juice from the spigot every two to three days. Left to accumulate, it can create excessive acidity. Dilute the drained juice at 1:100 with water and use it as a soil drench for garden beds, or dilute it at 1:200 and use it as a drain treatment to suppress odors in household drains. Both applications work well.

When the bin is full, seal it and leave it undisturbed for two weeks. Do not open it during this period. At the end of two weeks, the contents should smell pleasantly sour and fermented — like a mild vinegar or pickled vegetable product. White fungal growth on the surface is normal and beneficial. Blue, green, or black mold indicates air entered the bucket and the contents should be discarded and the process restarted.

Using the Output

Burying in the garden is the most direct approach. Dig a trench eight to ten inches deep, deposit the bokashi output, and backfill with soil. Mark the location and wait two to four weeks before planting in that spot. The fermented material breaks down rapidly in contact with soil organisms, and the ground at the burial site becomes noticeably darker and looser within a month.

If you have an active compost pile, the bokashi output goes directly into the center of it. The organic acids in the fermented material are a food source for compost bacteria, and the pile's heat will process the pre-compost within one to two weeks. Cover the addition with browns to prevent surface drying.

For houseplants and container gardens, the output can be worked into potting mix — but with one caveat. The fermented material must be diluted with a minimum two-to-one ratio of potting mix; applying too concentrated a dose in a closed container raises acidity enough to stress plant roots before the material fully neutralizes.

When I spent eight months in an apartment between houses, I ran a bokashi bin on the kitchen counter as my only composting option. I was skeptical going in — the pickled-food smell seemed like a reasonable price to pay, but I wasn't sure I was actually composting or just fermenting. When I eventually buried the output in a friend's garden and checked it three weeks later, the soil had transformed visibly. That afternoon I became a bokashi convert, which I remain.

Strengths, Limits, and When to Use It

The two-week fermentation cycle is genuinely fast for a first-stage process. A full bin sealed on Monday becomes usable bokashi pre-compost by the following Monday two weeks later. No other method handles meat and dairy in a domestic setting with comparable speed and odor control.

Two constraints are worth naming clearly. First, bokashi is a two-step process —
without the second step of soil incorporation or pile addition, the fermented output
has nowhere to go. Apartment composters who use bokashi need a reliable route to
a second stage, whether that's a community garden, a friend's pile, or large planter
boxes. Second, bokashi bran is an ongoing consumable cost. Households that make
their own bran eliminate this expense after the initial EM culture purchase, but
there is a learning curve.

Bokashi earns its place in kitchens where the waste stream runs wider than vegetables and fruit — households that cook meat, eat fish, generate dairy, and don't want
to maintain two separate input categories. It also suits any composting situation
where odor is a genuine constraint: a sealed bin in a cabinet produces far less
ambient smell than a worm bin running even slightly off-balance. Paired with a
worm bin, the two systems together handle the full range of household food waste
in under four square feet.

Chapter 16

Trench and Sheet Composting: Building Fertility in the Ground

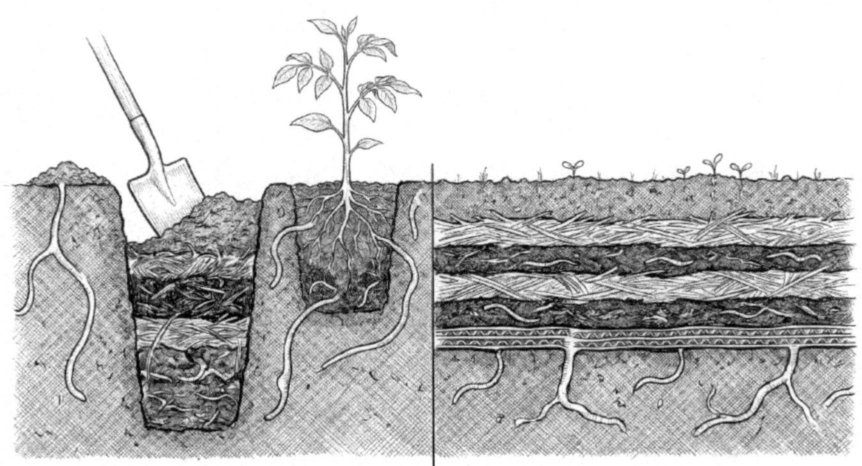

My grandmother never called it composting. She just called it "putting things back." Every autumn she'd dig a trench along the far edge of her vegetable plot and bury the season's spent plants, some kitchen scraps, and whatever fallen leaves she could reach without walking too far. By spring, that section of the garden was visibly darker and more productive than the rest. She was doing trench composting without ever having heard the term — and doing it well enough that I am still thinking about that garden forty years later.

Trench and sheet composting are the methods that build soil fertility without a bin, a turning schedule, or any container at all. The garden itself becomes the composting system. This is both their strength and their primary organizational challenge.

Trench Composting: Definition and Method

Trench composting means burying organic waste directly in the garden, in the ground where plants will eventually grow. It is among the oldest agricultural practices across cultures — Chinese, Japanese, Indigenous American, European — because it requires no infrastructure and builds fertility precisely where it is needed.

The standard trench is twelve inches wide and twelve inches deep. Dig it, add four to six inches of organic material — a mix of greens and browns if possible, or kitchen scraps alone if that's what you have — and backfill with the removed soil. The soil acts as a physical filter for odors and a direct inoculant of the buried material with the microorganisms needed to process it.

Mark the trench location. This matters more than it seems. A common problem with trench composting is digging into a buried trench before its contents have processed — which sets plant roots into unfinished material that may cause nitrogen theft or phytotoxicity. Stakes or flags indicating trench locations prevent this, as does a simple sketch of the garden with dates noted.

The dig-fill-plant rotation is the classic approach for small-to-medium vegetable gardens. Divide the garden into three zones. Year one: dig and fill zone A, grow in zone B, leave zone C fallow or planted in a cover crop. Year two: dig and fill zone B, grow in zone C, plant in zone A where last year's trench has fully decomposed. Year three: dig and fill zone C, grow in zone A, plant in zone B. The rotation cycles continuously, building soil fertility uniformly across the garden over three years.

Unfinished material in the trench breaks down in two to four months under most conditions, faster in summer, slower in cold soil. The trowel test works reliably:

probe the trench location with a hand trowel. If you can no longer distinguish the buried material from the surrounding soil, it has decomposed sufficiently for planting.

Trench Composting: Strengths and Limits

No bin, no pile, no turning, no dedicated composting area — the inputs go directly into the garden and stay there. This simplicity suits gardeners who do not want a separate composting operation or who garden in spaces where a visible pile would cause problems. The fertility builds in exactly the spot it will be used.

The constraints are those of any in-ground system: the garden layout needs to be planned to accommodate the rotation, and planting cannot happen immediately in trenched areas. For intensively planted small gardens where every square foot is in use continuously, the required fallow period creates a scheduling challenge. It also cannot handle large volumes rapidly — trench composting is a slow, continuous process rather than a batch one.

Sheet Composting: Patricia Lanza and the Lasagna Method

Sheet composting — also called lasagna gardening after Patricia Lanza's 1998 book of that name — is the approach of layering organic materials on the garden surface and allowing them to decompose in place. No digging, no turning, no soil preparation beyond what the decomposing layers themselves provide. New garden beds can be established over existing lawn, compacted soil, or even light gravel without any initial cultivation.

The foundational layer is cardboard or thick newspaper, overlapping at the edges to create a continuous barrier against existing vegetation. This layer serves two purposes: it smothers weeds and grass by excluding light, and it provides a high-carbon base layer that soil organisms begin decomposing from below. Wet the cardboard thoroughly before applying — dry cardboard repels water and undermines the whole system.

Above the cardboard, alternate layers of greens and browns exactly as you would in a conventional compost pile, but without turning. Browns twice as thick as greens

— typically four to six inches of browns for every two to three inches of greens. Continue layering until the bed reaches ten to fourteen inches high. It will settle significantly — half its initial height or more — as materials decompose over the following months.

Water the completed bed thoroughly. Every layer should feel damp throughout. Spread a layer of finished compost or topsoil on top if you intend to plant before full decomposition occurs — seedlings and transplants can be set into this surface layer while the layers below continue breaking down around and beneath their roots.

The Lasagna Bed That Worked Too Well

I built a lasagna garden one October using every scrap of cardboard I could find, six months of paper bags, autumn leaves, and three buckets of coffee grounds from the café down the street. By spring it had settled from fourteen inches to six, and what was underneath was the finest soil I had ever grown in. The tomatoes that summer were absurd. My neighbor — the skeptic from Book II — came over and looked at them for a long time without saying anything.

What that bed demonstrated is the no-till advantage of sheet composting: the layering process never disturbs the existing soil structure. Native earthworm populations remain intact. Fungal networks established over years in the underlying soil are undisturbed. The new organic material feeds that existing ecosystem from above rather than disrupting it.

Sheet Composting: Timeline and When to Choose It

Full decomposition of all layers takes six months to a year in temperate climates. A bed built in autumn is typically ready for spring planting if the layers were adequately moist and the green-to-brown ratio was balanced. Beds built in spring may not be fully ready until the following year — though planting through the layers into a surface layer of finished compost works in the interim.

Sheet composting excels at establishing new beds over challenging surfaces. It is the method of choice when starting a garden on lawn, converting a compacted or weedy area, or building beds in a location where digging is difficult. The material requirement is substantial — a ten-by-ten-foot bed fourteen inches high takes real volume — but autumn leaves, cardboard, and kitchen scraps accumulate to that volume over a season without special sourcing.

Choose trench composting if you have an existing garden with defined beds and a rotation is feasible. Choose sheet composting if you are building new growing areas, if no-till is a priority, or if you want to turn a lawn into a productive garden without the labor of excavation.

Chapter 17

Grasscycling and Green Waste Management

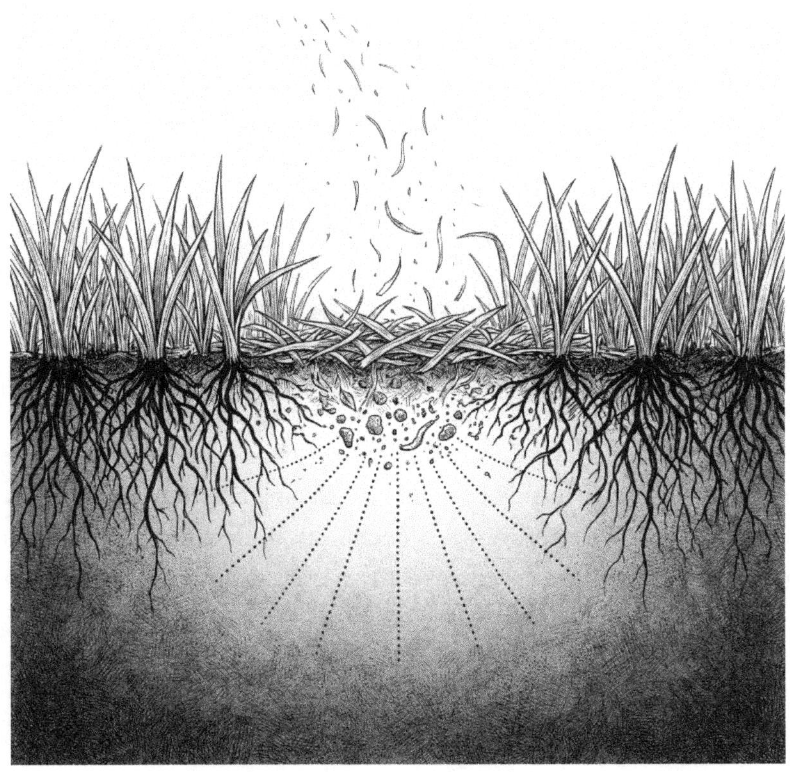

Ask most homeowners why they bag their lawn clippings and you will hear about thatch. The lawn will get matted. The clippings will pile up and block water. The grass will suffocate. It is one of the most durable pieces of gardening misinformation in circulation, and it persists because it sounds plausible. Thatch is stems and roots, not leaf blades. Leaving grass clippings on the lawn does not cause thatch. It returns approximately 25% of the lawn's annual nitrogen needs to the soil with zero effort.

That is grasscycling: mowing and leaving. No bagging, no composting, no additional steps. It is the simplest fertility-building practice in this entire book, and the argument against it turns out to be wrong.

How Grasscycling Works

Fresh grass clippings are approximately 80% water and break down rapidly on the lawn surface — typically within three to four days in warm weather. As they decompose, they release nitrogen, phosphorus, potassium, and trace minerals back into the soil. The organisms doing this work are the same soil bacteria and fungi already present in a healthy lawn, requiring no additional inputs.

The nitrogen return is meaningful. University extension research across multiple climates has consistently documented that grasscycled lawns require approximately one application less of nitrogen fertilizer per season than bagged lawns receiving the same general management. Over five years of grasscycling, the difference in fertilizer costs is measurable, and the difference in soil organic matter is observable.

Mulching mowers — mowers with a specially designed blade and deck that recirculates clippings and cuts them into finer particles before depositing them — accelerate decomposition relative to standard side-discharge mowers. They produce clippings small enough to work down between grass blades immediately rather than sitting on the surface. Any mower works for grasscycling; a mulching mower does it better.

The One-Third Rule and Timing

Never cut more than one-third of the blade length in a single mowing session. This rule matters for several reasons simultaneously. Removing more than one-third causes stress to the grass plant, which responds by diverting energy from root development to leaf regrowth. It also produces clippings that are too long and dense to break down quickly on the surface — these can clump and create temporary pockets of reduced air movement.

Mow when the grass is dry. Wet clippings clump together in dense masses rather than distributing evenly, and clumps large enough to smother the underlying grass do cause the kind of surface problems that grasscycling critics predict incorrectly for normal clippings. Mowing dry is the single most effective technique for consistent grasscycling results.

Vary your mowing direction or pattern from session to session. Mowing in the same direction repeatedly causes grass blades to lean permanently in that direction, reducing upright growth over time.

Integrating Clippings with Composting

Not all grass clippings should stay on the lawn. After heavy rain or rapid spring growth, excess clippings can accumulate faster than they decompose. These excess clippings — and clippings from a lawn recently treated with herbicides or pesticides — belong in the compost pile or in a separate green waste stream.

In a compost pile, fresh grass clippings are nitrogen-dense and decompose rapidly, but they compact quickly into anaerobic mats. Add them in layers no more than two inches thick, alternating with browns, or mix them throughout the pile during a turn rather than depositing them as a surface layer.

As mulch around garden plants, a thin layer of dried grass clippings — one to two inches, not fresh, not piled thickly — acts as a slow-release nitrogen source as it decomposes. Fresh clippings applied as mulch in thick layers generate heat during decomposition and can damage plant stems at the point of contact.

Managing Other Green Waste

Autumn leaves are the high-volume green waste challenge for most temperate-climate gardeners. The options are: add them to the active compost pile (shredded or mowed over), run a separate leaf mold operation, use them as mulch directly, or store them as a carbon reserve for the following year.

Leaf mold is worth a brief explanation because it is useful but distinct from standard compost. Pile leaves in a separate bin or heap, wet them, and leave them undisturbed for one to two years. The result is a dark, crumbly, earthy-smelling material that contains few nutrients but substantially improves soil structure and water retention. Leaf mold applied as mulch around established plants is an excellent use of material that would otherwise need to be composted or disposed of.

Pruning and woody material larger than a pencil's diameter breaks down slowly in a standard compost pile and is better chipped or shredded before adding. Unshredded woody prunings are appropriate for hugelkultur (Chapter 7) or as the coarse base layer of a new pile. A small electric chipper handles most domestic pruning volumes and substantially speeds the composting of woody material.

Diseased plant material is the case that requires judgment. If your compost pile consistently reaches 131°F throughout its mass and maintains that for three consecutive days, it can handle most plant diseases. Club root, fire blight, and similar soil-borne diseases require the full thermophilic treatment. If you are running a cold pile or are uncertain about temperatures, dispose of diseased material in the green waste bin or by burning where permitted, rather than cycling potential pathogens back through the garden.

Chapter 18

Hugelkultur: Composting with Wood and Buried Logs

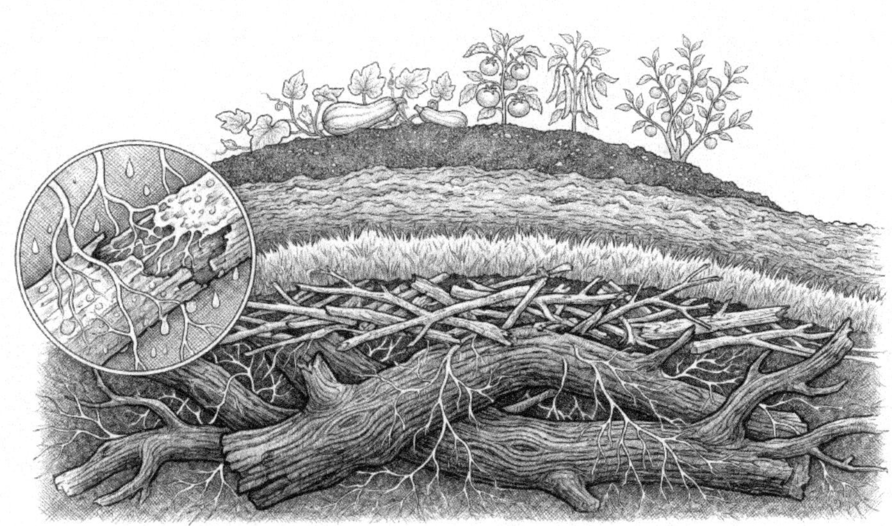

B ury a log. Wait five years. In the meantime, plant crops on top. That is the compressed description of hugelkultur — a Central European permaculture practice whose name translates roughly as "mound culture" — and it sounds more eccentric than it is. What happens inside a buried log over years is one of the more interesting long-arc fertility processes in gardening: the wood becomes a slow-release sponge, holding water, feeding fungi, and eventually decomposing into a self-fertilizing soil that requires progressively less management as it matures.

Hugelkultur is the only method in this book that builds a permanent raised bed as a byproduct of the composting process. The decomposing wood is not a component of the fertility system — it is the structural foundation of the bed itself.

The Science of Wood Decomposition

Wood decays through fungal action primarily. White rot fungi (Trametes, Ganoderma) break down both lignin and cellulose, leaving behind white, fibrous material. Brown rot fungi (Serpula, Coniophora) break down cellulose while leaving lignin intact, producing the brown, crumbly material characteristic of well-aged buried logs. Both processes are slow by the standards of bacterial composting — a buried hardwood log takes five to ten years to decompose fully — but that slowness is precisely what makes hugelkultur useful.

Decomposing wood absorbs and retains up to four times its dry weight in water. A log that weighed 20 pounds when buried may hold 80 pounds of water at peak moisture retention. This water-holding capacity peaks in years three to six of decomposition, when the wood has broken down enough to develop a porous, sponge-like structure. Hugelkultur beds built in drought-prone areas or on slopes that shed water perform noticeably better than conventional raised beds in dry conditions after the establishment period.

The nitrogen dynamics are the aspect of hugelkultur that most confuses first-time practitioners. Fresh wood, like any high-carbon material, draws nitrogen from its surroundings during initial decomposition. In years one and two, crops planted in a hugelkultur bed may show nitrogen deficiency symptoms even if the surrounding compost layer was generous. This is the establishment period, and it requires either nitrogen-tolerant crops or supplemental nitrogen inputs. By years three to five, as the wood begins releasing its stored nutrients, the beds become self-fertilizing and require little or no external input.

Building a Hugelkultur Bed

Site selection matters more in hugelkultur than in any other method because the bed is permanent. Orient the long axis to capture maximum rain runoff on sloped ground, or lay it perpendicular to prevailing winds on flat ground to minimize moisture loss. Ensure drainage is adequate — a hugelkultur bed that sits in standing water will not drain, and the anaerobic conditions created will kill the fungi that drive decomposition.

Wood selection is the decision that most affects long-term performance. Hardwoods — oak, apple, maple, alder, birch — are preferred for their density and slow decomposition rate. Softwoods (pine, spruce) work but decompose faster and produce a more acidic environment. Several species should be avoided: black walnut produces juglone, a compound toxic to many plants; cedar and black locust contain natural preservatives that resist decomposition and may inhibit soil organisms. Fresh-cut wood can be used but will draw more nitrogen in the establishment phase; partially rotted wood — the soft, punky material from a log that has been sitting in the garden for a year — is ideal because the decomposition process has already begun.

The layer sequence from base to surface: largest logs go at the bottom, filling the lowest third of the bed's depth. Above these, smaller branches and woody debris. Then a layer of sod or grass cut upside down — roots and soil facing up, blades down. Above that, a generous layer of finished compost or aged manure. Finally, topsoil to a depth of four to six inches. The bed at completion stands two to four feet high; expect it to settle to half that height within the first year as materials compress and begin decomposing.

Planting, Management, and Maturity

In year one, plant crops that fix nitrogen or are tolerant of lower nitrogen availability: beans, peas, nasturtiums, squash, and most brassicas perform adequately during the establishment phase. Avoid heavy nitrogen feeders — corn, tomatoes, leafy greens requiring lush growth — until year two or three when the nitrogen dynamics stabilize.

Irrigation in year one is closer to a standard bed: water regularly while the wood establishes its moisture-retention capacity. By years two to three, a hugelkultur bed typically needs 30 to 50% less irrigation than a conventional bed of the same size in the same conditions. By years five and beyond, some practitioners in humid climates report that mature hugelkultur beds need supplemental irrigation only during the most severe drought.

The connection to fungal-dominated compost is direct and worth noting for gardeners who have read Book I Chapter 2. A hugelkultur bed is, over its life, an increasingly fungal-dominated system. The network of fungal mycelium that develops around the decomposing wood creates ideal conditions for perennial vegetables, fruit bushes, berries, and dwarf fruit trees. The beds work adequately for annual vegetables in the establishment years, but their long-term highest expression is as perennial production systems.

Strengths, Constraints, and When to Use It

A well-built hugelkultur bed produces fertility from materials that would otherwise need to be chipped and composted separately, requires progressively less intervention as it matures, and in areas with poor or depleted soil offers a route to productive growing that does not depend on large volumes of imported compost. In drought-prone areas, the water retention alone justifies the initial construction effort for established gardeners.

The construction is physically demanding and space-intensive — a proper hugelkultur bed is not suitable for a 20-square-foot balcony garden. The nitrogen drawdown in the establishment years requires planning. And the bed is permanent: once built, moving it means dismantling years of developing decomposer communities. Choose this method for larger gardens, sloping terrain, areas with low or irregular rainfall, and situations where woody waste that would otherwise require disposal can be incorporated productively.

PART IV

COMPOSTING WITHOUT A BACKYARD

Urban, Indoor,
and Community Solutions

Chapter 19

The Urban Composter's Mindset: Constraints as Opportunities

Every urban composter faces a version of the same conversation. Someone mentions they don't compost because they live in an apartment, or don't have a garden, or the building doesn't allow it, or there's nowhere to put a pile. Behind each of those reasons is an implicit assumption: that composting requires a yard. It doesn't. It requires a food waste stream — which every household has — and a system sized to the space and conditions available. The constraints of urban life are a design brief, not a disqualification.

This book's preceding chapters covered methods that assume outdoor space. Books IV covers what replaces them — or supplements them — when that space is absent or severely limited. The methods themselves are not compromises. Vermicomposting, bokashi, and small-scale tumbler composting in urban settings produce outputs comparable in quality to anything covered in Book III. The volume is smaller. The footprint is smaller. The output per unit of input is, in some cases, higher.

How Urban Composting Differs in Practice

The differences are operational, not biological. Decomposition follows the same chemistry on a kitchen counter as in a suburban backyard. What changes is the tolerance for odor, the proximity of neighbors, the absence of outdoor space to absorb liquids or process bulk material, and the design requirements that follow from all three.

Urban composting is, almost by definition, managed more tightly. An open pile in a backyard can absorb minor management errors — a period of neglect, a wet autumn, an oversupply of kitchen scraps — without immediate consequences. An indoor worm bin or a kitchen bokashi bucket has less margin. The smaller the system, the more sensitive it is to imbalance, and the more attentive management it requires to stay in good working order.

That attentiveness need not translate to significant time investment. A well-established worm bin requires five to ten minutes of attention per week. A bokashi system requires two to three minutes at each feeding and a ten-second juice drain every few days. The learning curve is the intensive part; the maintenance, once the system is running well, is genuinely minimal.

Choosing the Right Urban System

Five factors guide the selection: available space measured in square feet or linear feet of counter or cabinet, weekly food waste volume in pounds, sensitivity to odor in a shared living environment, budget for initial setup, and whether outdoor access (community garden, planter boxes, shared building courtyard) exists to receive processed output.

Vermicomposting needs a surface area of one square foot per pound of weekly food waste, tolerates most building temperatures, produces minimal odor when correctly managed, and generates castings that can be used in houseplants or balcony containers. It is the system with the broadest applicability across urban living situations.

Bokashi requires only the footprint of one or two five-gallon buckets, handles meat and dairy that worms will not, produces minimal odor from the bin itself, and requires no temperature management. The constraint is the two-step process: the fermented output needs to go somewhere to complete decomposition, which requires either outdoor access or supplementary system.

Electric countertop composters require only counter or cabinet space, no outdoor access, and minimal ongoing management. They reduce food waste volume significantly and produce a stable, odor-neutral output. They are the right choice for households with no outdoor access at all and no appetite for live worm management.

Community composting programs — drop-off points, building-level programs, pickup services — are the option for households that want to divert food waste but cannot manage an in-home system. They require no equipment and no ongoing management beyond sorting and transporting waste to a collection point.

Odor Management

Composting odors in an urban setting come from two sources: decomposing material exposed to air before it reaches the composting system, and composting systems that are out of balance. The first is managed by a countertop collection container with a carbon filter, emptied every two to three days. The second is managed by correctly running the system.

A balanced worm bin smells like garden soil. A balanced bokashi bin, properly sealed, smells like nothing from the outside. An electric composter with a functioning carbon filter produces no detectable odor during operation. If any of these systems smell bad, the smell is diagnostic: the system needs attention, not acceptance.

The chemistry is simple: odors from composting organic matter are either ammonia-based (too much nitrogen, aerobic) or hydrogen sulfide and methane-based (anaerobic). Both conditions have specific causes and specific fixes covered in the individual method chapters. In an urban context where odor affects neighbors or other building residents, identifying and resolving the cause quickly matters more than it does in an isolated backyard setting.

Pest Management in Dense Living Situations

In a detached house, a fruit fly outbreak in the kitchen compost pail is an annoyance. In a shared apartment building, it is a neighbor relation problem. Pest prevention in urban composting systems requires sealed containers, not just covered ones, and consistent management of inputs.

Fruit flies breed in exposed organic matter, not in closed systems. A properly sealed worm bin with food buried under bedding, or a bokashi bin with a tight-fitting lid, does not generate fruit flies. The source of most urban composting pest problems is the collection stage — scraps sitting in an open bowl or a loosely lidded countertop container for days at a time, particularly in summer.

Rodents and larger pests are almost never a problem with indoor systems. They become relevant only when composting moves outdoors to a balcony, courtyard, or shared building space — situations addressed in Chapter 3 and Chapter 5.

Chapter 20

Indoor Systems: Worm Bins, Bokashi Buckets, and Electric Composters

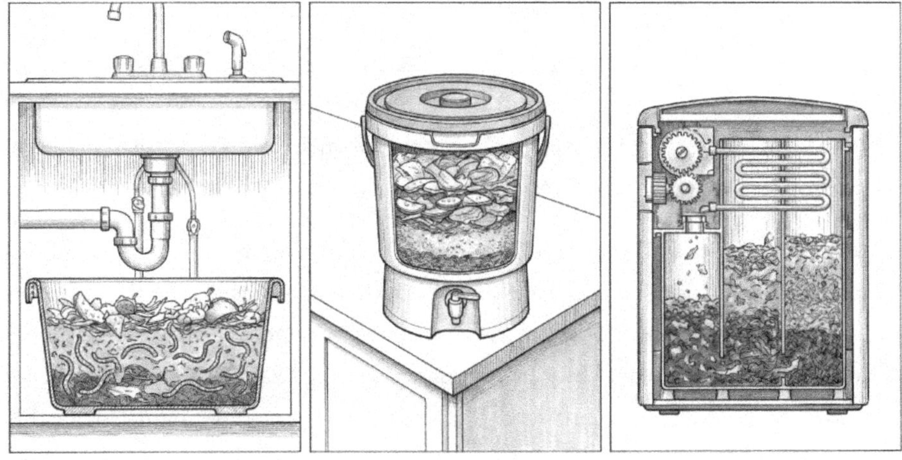

The space under a kitchen sink is 18 inches wide, 18 inches deep, and 24 inches high in most standard cabinets. That is enough room for a five-gallon worm bin on one side and a bokashi bucket on the other — a two-system setup that handles the complete range of household food waste, from fruit peels and coffee grounds to the chicken bones and leftover curry that a worm bin cannot process.

Many people composting successfully in apartments are doing it in roughly that space, and have been for years.

The technical details of running each method — materials, schedules, troubleshooting — are in Book III. Here, the focus is the apartment and indoor adaptation: placement, temperature management, odor control, and how to handle the output when outdoor access is limited.

Vermicomposting Indoors

Worm bins placed indoors are protected from the temperature extremes that affect outdoor systems. Red wigglers prefer 55°F to 77°F (13°C to 25°C) — a range that matches typical indoor environments almost exactly. A bin placed under the kitchen sink or in a kitchen cabinet maintains appropriate temperature year-round in most heated dwellings without any intervention.

Basements and storage areas are workable locations if temperatures stay within range. Avoid placement near heat sources — radiators, ovens, dishwashers — which can raise local temperature above 85°F and stress worms. Avoid uninsulated garages or balcony storage where temperatures drop below 40°F in winter.

Moisture management is more critical indoors than outdoors. Without natural air circulation, an indoor bin can accumulate condensation inside the lid, dripping onto the bedding surface and raising moisture levels beyond the target range. Check for condensation weekly and blot or ventilate as needed. Adding a layer of dry torn cardboard to the surface absorbs excess moisture and provides a light barrier that discourages fruit flies.

An apartment-scale system — a single 10-gallon bin — handles two to three pounds of weekly food waste. Commercial stackable systems (Urban Worm Bag, Can-O-Worms, Worm Factory) manage higher volumes in the same footprint, using stacked trays that allow continuous feeding and straightforward harvesting without the dump-and-sort process. For households generating four to six pounds per week, a stackable system is worth the additional cost.

Bokashi Indoors

A kitchen bokashi setup needs three things positioned conveniently: the active bin for daily additions, a storage location for completed full bins during the two-week fermentation period, and a place to drain juice every few days. Most household kitchens can accommodate this within a cabinet or pantry space.

The fermentation smell from a properly sealed bokashi bin is negligible at the exterior — the airtight lid prevents escape of the sour fermentation odors inside. Opening the bin during the two-week rest period releases a noticeable sour smell for the seconds the lid is off. Outside that, a sealed bokashi bin in a kitchen cupboard produces no odor detectable from across the room.

The bokashi juice — drained from the spigot every two to three days — is best disposed of by diluting and pouring down the kitchen drain or toilet. At 1:100 dilution with water, it acts as a drain treatment, breaking down grease accumulation in the drain. For apartment dwellers without outdoor soil access, this is often the most practical routine disposal method.

Using the fermented output without outdoor access requires a secondary system. Options: adding it to a community garden compost pile (contact your local community garden — most welcome additions from known sources), mixing in small quantities into large container planters or window boxes at a 1:5 ratio with potting mix, or finding a building resident or nearby friend with garden access who accepts it regularly. Some neighborhoods have informal networks for exactly this purpose.

Electric Composters

Electric countertop composters are the indoor option that requires the least tolerance for biological processes. No worms, no fermentation smells, no living organisms to manage. The machine processes food waste through a programmed cycle of heat, aeration, and mechanical action over four to eight hours, reducing volume by 70 to 90% and producing a dry, granular output.

Current leading models include the Lomi (Pela), the Vitamix FoodCycler, and the Reencle, each with different capacity, cycle times, energy use, and approach to drying versus true composting. The Reencle uses a microbial culture to begin active decomposition, producing an output closer to pre-compost. The others primarily dehydrate and grind, producing a dry food waste powder that is stable for storage but requires soil contact to complete breakdown.

The honest evaluation for apartment use: electric composters are the most convenient indoor option for households that want to divert food waste and are not interested in managing a biological system. The output is genuinely useful as a soil amendment. The electricity cost and purchase price are the trade-offs. For a household paying market rates for organic compost amendments to houseplants or a small balcony garden, the economics are reasonable over two to three years.

The Two-System Approach

The most complete indoor composting setup combines a worm bin for plant-based kitchen waste — the majority of what most households generate — with a bokashi bin for meat, dairy, fish, and cooked foods. Each system handles what the other cannot.

In practice, the workflow is straightforward. Vegetable peels, fruit scraps, coffee grounds, eggshells, and paper go into the worm bin. Meat scraps, cooked leftovers, dairy, and citrus in quantity go into the bokashi bin. The worm bin produces castings every two to three months. The bokashi bin produces pre-compost every two weeks.

The total footprint is under four square feet for a household generating four to six pounds of weekly food waste. The total active management time is under fifteen minutes per week. Two systems running in parallel divert close to 100% of household food waste, which is a more complete outcome than any single indoor system achieves.

Chapter 21

Balcony and Small-Space Composting

A standard urban balcony is sixty square feet. Most are smaller. Some building regulations specify that composting equipment on a balcony must not be visible from the street. Some buildings prohibit it entirely. These are the actual conditions of balcony composting, and working from them produces better outcomes than pretending the constraints are less significant than they are.

Within those constraints, meaningful composting is possible. What changes relative to a backyard setup is scale, container selection, and the degree to which every decision needs to account for neighbors, structure, and aesthetics.

Weight and Drainage

Before placing any composting system on a balcony, check the structural load rating for the floor. Most urban balconies are rated for 40 to 60 pounds per square foot. A full compost tumbler holding 50 gallons of material weighs approximately 400 to 500 pounds — ten pounds per square foot for a tumbler occupying a four-square-foot footprint. Distributed carefully, this is within typical ratings; concentrated in one corner, it may exceed them.

Drainage is the second structural concern. A worm bin or compost tumbler that leaks liquid onto a balcony floor can stain concrete, damage the waterproofing layer, and cause problems for neighbors in lower-floor units. Use systems with built-in trays or place collection trays under any system with drainage. Empty collection trays regularly.

System Options for Balconies

Bokashi is the most consistently balcony-compatible system. The bin is sealed and produces no odors at the exterior, generates no leakage if drained regularly, requires minimal floor space, and involves no living organisms whose welfare depends on outdoor conditions. A bokashi bin on an enclosed balcony performs identically to one in a kitchen.

Mini stackable worm towers designed specifically for balcony use have a footprint of roughly 18 by 18 inches and process two to three pounds of weekly food waste. They should be positioned away from direct summer sun, which can raise temperatures inside the units above 85°F and stress worm populations. A shaded balcony corner or a position against a north-facing wall is appropriate.

Compost tumblers sized for balcony use — 20 to 30 gallon models — exist, though the selection is narrower than for garden-scale units. Look for models with a sealed base tray, a dual-chamber design, and a frame that keeps the drum off the floor surface. These produce genuinely hot compost in small batches, though the volume constraints make them practical only for households with modest waste streams.

Aesthetics and Neighbor Relations

A composting system that looks intentional is easier to defend to a skeptical building manager than one that looks improvised. Several manufacturers produce worm bins and bokashi systems in designs that could pass for kitchen appliances or planters. The Worm Farm series, the Bokashi One, and comparable products prioritize appearance without compromising function.

Screening an outdoor balcony system with a bamboo or trellis panel serves two purposes: it provides partial shade that benefits the composting organisms, and it makes the system invisible from neighboring balconies and from the street. A planter box with trailing plants on the screening panel removes any aesthetic friction entirely.

Regulations and Building Permissions

Check your lease and building rules before setting up any composting system on a shared balcony or in a shared outdoor space. Many leases are silent on composting; others prohibit food waste storage on balconies specifically. Where rules are silent, a brief, factual conversation with building management — framing the system as a sealed container for food waste processing, which produces no odor and no liquid runoff — is generally more productive than asking for permission and receiving a preemptive no.

For buildings where composting is currently prohibited, the most effective approach is proposing a pilot: one resident, one sealed system, three months. Document the absence of odor complaints or pest issues. Use that record as the basis for a building-wide program proposal. This is how most successful building composting programs begin.

HOA rules on composting vary widely by jurisdiction and by individual HOA. Some states — California, Colorado, New York among them — have passed laws restricting HOA authority to prohibit composting outright. Knowing your jurisdiction's position before any conflict arises is worthwhile.

Chapter 22

Regulations, Incentives, and the Legal Landscape

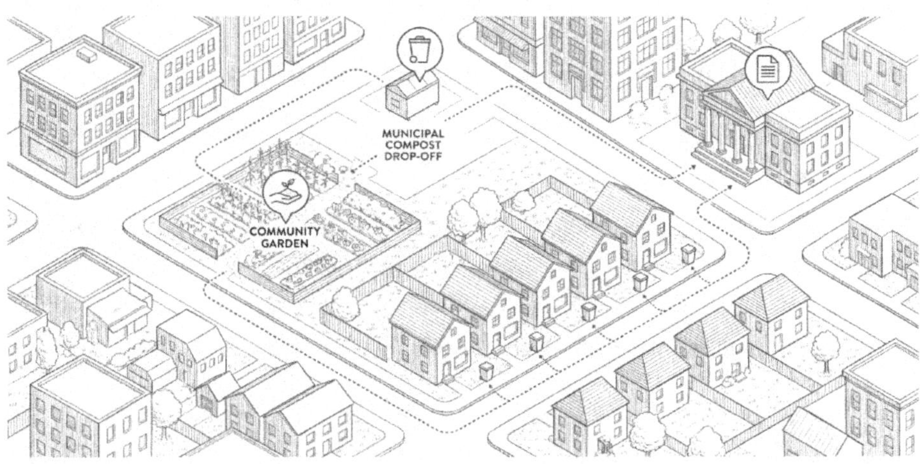

I n San Francisco, failing to separate food scraps from landfill-bound trash can result in a fine. In Vermont, food scraps have been banned from disposal as landfill since 2020 for any generator producing more than 40 pounds per week. In parts of New York City, certain building sizes are now required by law to compost or subscribe to organics collection. These laws exist, are enforced, and affect people who have never thought about the legal dimension of what goes in their garbage bag.

Most composters come to the practice through interest in gardening or sustainability, not through regulatory obligation. But the legal and financial landscape around composting has changed enough in recent years that understanding it has practical value — both to avoid unintentional violations and to claim incentives that most eligible households never access.

Municipal Composting Mandates

California Senate Bill 1383, which took effect in 2022, requires that all California residents have access to organic waste collection service and that most residents participate. It is the most sweeping mandatory organics composting law in the United States, and it affects roughly 40 million people. Seattle, San Francisco, and several other cities implemented mandatory composting programs years earlier, building the regulatory and logistical infrastructure that informed the California law.

What mandatory composting means in practice varies by municipality. In some cities it means sorting food scraps into a designated bin for collection. In others it includes home composting as a compliance option, sometimes with a bin subsidy attached. In Vermont, the system relies on commercial composting infrastructure for urban areas and home composting for rural residents.

The trend is toward expansion. As of 2024, fifteen US states have passed or are actively considering some form of organics diversion requirement. The European Union's Waste Framework Directive requires member states to have separate food waste collection for all households by 2025. Understanding whether your jurisdiction has existing or pending requirements takes a five-minute search on your municipal waste management website.

Financial Incentives for Home Composting

Subsidized or free compost bins are available from municipal programs in hundreds of US cities and counties. The programs are often poorly publicized and significantly underutilized. Residents in qualifying areas can receive a plastic compost bin, a tumbler, or in some cases a worm bin at no cost or at 50 to 75% subsidy simply by applying through their local solid waste authority. The National Resources Defense Council estimates that fewer than 15% of eligible residents in active bin subsidy programs have claimed their bins.

Waste disposal fee reductions exist in municipalities operating pay-as-you-throw programs, where residents pay for waste collection based on volume or weight of landfill-bound trash. Households that compost — and therefore generate less landfill waste — qualify for reduced collection fees in these systems. The annual savings range from $30 to $120 depending on the municipality and the volume diverted.

Tax incentives for composting are less common than bin subsidies but exist in specific jurisdictions, particularly for community composting operations, school programs, and businesses. Several states offer income tax credits or deductions for the purchase of composting equipment, and some counties offer reduced property tax rates to landowners maintaining active composting programs as part of agricultural preservation easements.

What Is Legal to Compost on Your Property

For standard backyard composting of plant-based food waste and yard material, virtually every US jurisdiction permits it without restriction. The legal complexity begins with specific inputs — manure, meat and dairy, biosolids — and with proximity to water sources, property lines, and structures.

Many municipalities prohibit composting meat and dairy in residential zones, regardless of the method. The prohibition is typically aimed at open pile composting and pest attraction rather than sealed systems like bokashi. Whether a sealed, odor-free bokashi bucket technically falls under a "no meat composting" restriction is genuinely ambiguous in most municipal codes, and the practical risk of enforcement is low. If your jurisdiction prohibits it and you choose to use bokashi for these materials, the risk is yours to assess.

HOA restrictions on composting have been explicitly limited in California (AB 1572, 2021), Colorado, and several other states. If you live in a jurisdiction with

HOA composting protection laws, those laws override HOA rules that prohibit composting. Knowing this before an HOA dispute arises is considerably more useful than discovering it during one.

Finding Your Local Resources

The most reliable starting points are your local municipality's solid waste or environmental services department, and your state's department of environmental quality or natural resources. Both typically maintain pages on composting programs, bin subsidies, and disposal requirements.

The US Composting Council (www.compostingcouncil.org) maintains a searchable database of composting programs, facilities, and industry standards. The EPA's food recovery hierarchy provides context for where composting sits within the broader waste reduction priority framework — below prevention and redistribution, but above landfilling and incineration.

State land grant university extension services are among the most practically useful resources for composting information specific to local climate, materials, and regulations. Extension publications are peer-reviewed, regionally adapted, and freely available online in every state.

Chapter 23

Community Composting: Collective Action, Shared Benefits

On a Saturday morning in late April, the community garden at 7th and Oak fills by eight o'clock. Plots are being turned. Children are being handed seed packets. Someone has brought a flat of tomato transplants to share. At the back corner, three large compost bins receive the week's accumulated yard waste from residents who brought it in bags, and the finished compost from the adjacent bins is being loaded into wheelbarrows by the people waiting to amend their plots.

Nothing about this arrangement required a policy document. It required a site, some bins, a loose schedule, and enough people who cared about the same problem to build something that worked.

Community composting, at its best, is that kind of self-organizing mutual infrastructure. At its worst, it is a poorly managed shared pile that becomes a source of neighborhood conflict. The difference between the two is almost entirely organizational.

Models of Community Composting

Drop-off points and collection programs are the simplest model: residents bring materials to a central location, and a managing organization handles processing. These work well in high-density areas where individual composting is impractical and where a central processing site — a community garden, a school with land, a church with a parking lot — is accessible to a walkable catchment area.

Shared composting bins in community gardens operate differently: multiple gardeners contribute to shared piles and share the output. This model creates the clearest incentive alignment — everyone who contributes gets compost — and tends to generate strong voluntary engagement. It works less well in gardens with significant membership turnover or where contribution monitoring creates conflict.

Pickup composting services — commercial and non-profit — collect food scraps from subscribers on a weekly or biweekly basis, process them at an off-site facility, and typically offer subscribers a portion of the finished compost in return. These services have expanded significantly in urban areas over the past decade and provide the most complete urban composting solution for households without outdoor access and without interest in managing a home system.

City-wide collection and industrial processing programs — the San Francisco and Seattle models discussed in Chapter 6 — are the most scalable version, treating organics diversion as municipal infrastructure rather than voluntary participation. These programs are covered in detail in the next chapter.

Starting a Community Composting Initiative

The gap between "someone should do this" and "this exists" in community composting is almost always organizational capacity, not material resources. Bins can be obtained cheaply or free through municipal programs. Space can usually be found through schools, faith communities, or parks departments. What requires the most sustained effort is building a team that can manage ongoing operations through the inevitable turnover of founding members.

Before acquiring anything, spend time on site selection. The site needs adequate drainage — standing water in a composting area creates anaerobic conditions and persistent odor. It needs vehicle access if bulk material deliveries or compost distribution are planned. It needs distance from residential windows sufficient that odor from a momentarily imbalanced pile does not generate complaints. And it needs a formal agreement — in writing — with whoever controls the land.

Equipment procurement follows site establishment. For a program serving 50 to 100 households, a three-bin turning system — each bin approximately one cubic yard — is a practical starting point. Municipal solid waste departments frequently supply bins at subsidized cost or free for community programs. Pitchforks, a compost thermometer, a moisture meter, and a wheelbarrow complete the basic toolkit.

Operations need a written protocol simple enough that volunteers can follow it without prior composting experience: which materials are accepted, how to layer additions, the turning schedule, how finished compost is distributed, who to contact with problems. Contamination — non-compostable materials appearing in the drop-off stream — is the most common operational problem. Labeled bins, visual guides posted prominently, and a designated material checker at drop-off points reduce contamination more effectively than any other single measure.

Funding, Volunteer Retention, and Legal Considerations

Community composting programs typically fund their initial setup through a combination of municipal grants (the EPA and USDA both offer relevant programs), small business or local organization sponsorships, and volunteer labor. Ongoing operations are funded through compost sales, membership fees, and in some cases municipal contracts for organics diversion services.

Volunteer retention is the factor most commonly underestimated in program planning. The people who start a composting program are rarely the same people who maintain it three years later. Building retention requires recognizing contribution visibly, creating skill-building opportunities that deepen engagement, and distributing compost generously to active contributors — the direct, tangible reward that keeps people returning to a muddy pile on a rainy Saturday.

Liability for community composting operations varies by organizational structure and jurisdiction. Programs operated under the umbrella of an existing non-profit or school typically carry liability coverage through the parent organization. Standalone operations should consult a local attorney about appropriate organizational structure — a simple unincorporated association may suffice for a small neighborhood program, while a larger operation serving multiple schools or hundreds of households is better structured as a formal non-profit.

Chapter 24

Urban Composting Around the World: Lessons from the Field

The composting literature is full of programs that worked under ideal conditions, with adequate funding, engaged participants, and supportive policy. Case studies from cities that have run large-scale organics programs for a decade or more offer something more useful: evidence of what survives contact with operational reality, funding gaps, political changes, and the full variability of a large urban population's composting compliance.

Four programs worth examining in detail: San Francisco for scale and policy structure, Seattle for enforcement approach, New York City for the specific challenges of density, and Todmorden in England for the grassroots model that works without top-down infrastructure. Each teaches something different.

San Francisco

San Francisco implemented mandatory composting in 2009, requiring all residents and businesses to sort waste into three streams: landfill (black bin), recycling (blue bin), and compost (green bin). The city chose a contractor model: Recology, a San Francisco-based waste management company, handles collection and processing at the Jepson Prairie Organics facility in Vacaville.

The scale is instructive. Jepson Prairie processes roughly 400 tons of organic material per day, producing approximately 50,000 tons of compost annually. A significant portion is sold to Napa and Sonoma Valley vineyards, creating a commercial market that makes the operation economically self-sustaining — an outcome that many composting programs aspire to and few achieve at comparable scale.

San Francisco's landfill diversion rate reached over 80% within five years of the mandatory program's implementation, compared to roughly 50% under the voluntary system that preceded it. The lesson is straightforward: voluntary programs plateau. The incremental gains from adding more outreach to a voluntary system are small. Mandatory programs with enforcement — initially fines for repeat violations, but more effectively social norm enforcement — produce step-change improvements in participation rates.

The contamination challenge is persistent. Plastic bags, non-compostable "compostable" serviceware, and general confusion about what belongs in the green bin remain ongoing issues a decade after implementation. Recology's solution is ongoing visual education at the point of collection — stickers, flyers, and direct feedback to high-contamination accounts — rather than punitive enforcement for first-time or occasional errors.

Seattle

Seattle has required food scraps to be placed in yard waste collection bins since 2015 for all single-family residences. The enforcement mechanism is a combination of inspection — waste collectors flag bins with visible food contamination — and financial penalties for persistent violators, ranging from a warning to a $50 fine for repeated non-compliance.

Cedar Grove Composting, which processes the material, operates one of the largest windrow composting operations in the Pacific Northwest. The finished compost — certified organic — is sold commercially and has been used extensively in regional landscaping and restoration projects, including several Washington State Department of Transportation highway slope stabilization projects where compost application proved more effective than conventional erosion control at a lower long-term cost.

Seattle's experience with the rollout of enforcement is worth noting for anyone designing a similar program. Immediate strict enforcement generated significant public resistance in the first year. Phased enforcement — warnings before fines, fines before service suspension — with substantial public education investment in the transition period produced better long-term compliance than early punitive approaches.

New York City

New York City's situation is instructive precisely because of its difficulty. Density, diverse languages, renter-dominated housing, and the logistical complexity of collection in a city of 8 million people make implementation substantially harder than in San Francisco or Seattle. The NYC Compost Project has operated since 1993 through a network of neighborhood drop-off sites, school composting programs, and community composting hubs — a decentralized model suited to the city's scale and heterogeneity.

GrowNYC's greenmarket food scrap collection network has been particularly effective. By embedding food scrap collection at existing farmers markets — weekly events that already draw environmentally engaged urban consumers — the program achieved high participation rates without requiring residents to change their daily routes. The co-location insight is transferable: composting participation increases when the drop-off point is at a location people already visit regularly for other reasons.

New York City began mandatory residential composting in 2024, following years of voluntary programs and several pilot mandatory programs in specific districts. The infrastructure challenge — providing collection bins and servicing them across thousands of dense city blocks — required several years of phased rollout. The city's experience demonstrates that mandatory programs require investment in collection infrastructure before enforcement, not simultaneously.

Todmorden, England

The Incredible Edible Todmorden project, begun in 2008 in a small mill town in West Yorkshire, took a different approach to both composting and urban food production. Rather than building a formal program with infrastructure and funding, a small group of residents began planting edible crops in public spaces — on road verges, outside the police station, along the canal towpath — and inviting the community to harvest freely.

Community composting became a supporting system for those public beds, fed by resident contributions and producing amendments for the plantings. The program has no formal infrastructure budget, no paid staff, and no enforcement mechanism. It runs on voluntary contribution maintained by visible, accessible benefit: anyone who contributes to the compost can eat from the gardens.

What Todmorden demonstrates is that the social infrastructure of composting — the mutual obligation, the shared purpose, the visible return — can drive participation without policy mandates or financial incentives when the benefit is immediate and the barrier to participation is genuinely low. It is not scalable to eight million people. But as a model for neighborhoods and small communities where informal social bonds are strong, it remains one of the most durable grassroots composting programs operating anywhere.

Transferable Lessons

Four conclusions emerge consistently across these programs. First, education before enforcement: mandatory programs that begin with education campaigns and grace periods achieve higher steady-state compliance than those that lead with penalties.

Second, simplicity of participation: the fewer decisions a participant must make at the point of contribution, the higher the participation rate. Single-stream collection with sorting at the processing facility outperforms complex multi-stream sorting at the household level in dense urban environments.

Third, visible local benefit closes the loop that makes composting feel worthwhile rather than merely obligatory. When residents can see that their food scraps become compost that goes to local parks, school gardens, or urban farms, participation is more durable than when the output disappears into an abstract municipal process. Fourth, co-location of collection points at existing destinations — markets, transit stations, schools, faith communities — consistently outperforms standalone collection infrastructure in voluntary participation models.

For urban composters anywhere in the world, the practical takeaway from these programs is this: what you do at home is the smallest unit of a larger system that is being built, unevenly and imperfectly, in cities around the world. Knowing how that system works — and what you can advocate for in your own city — is part of what it means to compost with full awareness of its context.

PART V

BLACK GOLD FOR YOUR GARDEN

Soil Science, Application, and Plant Health

Chapter 25

Understanding Your Soil: Types, Testing, and What Compost Changes

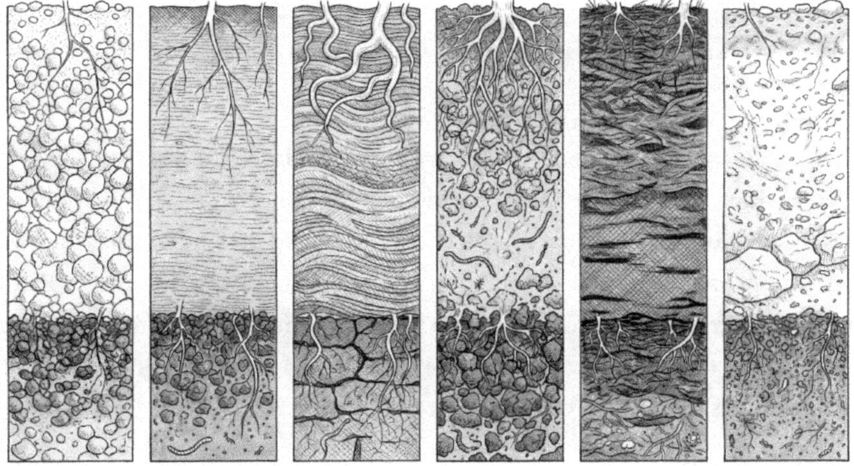

The first year I gardened on degraded soil, I planted a row of tomatoes and watched them sulk for three months. The plants were technically alive. They produced a handful of small, pale fruit in October and then gave up. The soil was compacted clay with almost no organic matter — the kind of ground that bakes hard in summer and sits waterlogged through spring. I added compost the following autumn. By the second year, the same bed produced more tomatoes than I could give away. I have been somewhat obsessive about soil ever since.

Compost works differently depending on the soil it enters. Understanding what kind of soil you have is not a bureaucratic prerequisite — it is the step that determines where your compost does the most good and how much you need to apply. What follows is the framework for making that assessment.

The Six Soil Types

Sandy soil sits at one end of the texture spectrum. It drains fast, warms quickly in spring, and is easy to work, but it holds neither water nor nutrients well. Compost improves it through aggregate formation — binding sand particles together into larger clusters that trap moisture and create sites where nutrients can attach rather than drain straight through. Two to three inches of incorporated compost per year, sustained over several seasons, transforms the water-holding capacity of a sandy bed measurably.

Clay is the opposite problem. The particles are so fine and flat that they pack together almost without air space, leaving drainage poor and workability low — clay soil dries into a kind of ceramic and becomes sticky and dense when wet. Compost opens this structure by creating aggregates that separate the clay particles, introducing permanent pore spaces. The improvement here tends to be faster and more dramatic than in sandy soil, because clay is already nutrient-rich; it just needs better physical structure to release what it holds.

Silty soil falls between the two in particle size, drains reasonably, and is often fertile, but shares clay's tendency to compact under foot traffic or heavy rain, forming a surface crust that sheds water rather than absorbing it. A consistent compost top dressing — two inches per year — maintains the loose, crumbly surface tilth that prevents this.

Loamy soil is the gardener's ideal: a blend of sand, silt, and clay particles in roughly equal proportion, well-draining but moisture-retentive, easy to work, and biologically active. Compost maintains rather than transforms it. Annual applications replenish organic matter consumed by microbial activity and plant uptake, keeping the balance that makes loam productive.

Peaty soils carry high organic matter already but tend toward acidity and can be waterlogged. Compost application here is targeted rather than corrective across the board: well-aged compost from a broad material mix gradually raises pH, while the physical structure it adds improves drainage in waterlogged conditions. Overdoing it risks tipping an already organically rich soil toward nutrient imbalance.

Chalky and alkaline soils present a different challenge — high pH locks out iron, manganese, and boron regardless of whether they are physically present, producing yellowing plants even in apparently fertile ground. Compost buffers pH from both directions: the organic acids produced during decomposition gradually lower alkaline pH, and the cation exchange capacity it builds helps moderate fluctuations. The correction is slow but cumulative.

Testing Your Soil: DIY Methods

The jar test reveals texture without any equipment beyond a glass jar, water, and the soil itself. Fill the jar one-third with soil, top it with water, cap and shake vigorously, then leave it undisturbed for 24 hours. Sand settles within the first two minutes; silt settles over the following hour or two; clay stays suspended longest, forming the top layer. The relative depth of each layer shows you the approximate composition of your soil. It is not laboratory-precise, but it answers the practical question accurately enough.

Drainage is tested by digging a hole roughly 12 inches deep and 12 inches across, filling it with water, and watching. If it drains in under an hour, drainage is fast — you are likely dealing with sandy or very loamy soil. If it takes two to three hours, drainage is moderate and normal. If standing water remains after six hours, drainage is poor enough to require either structural intervention or adaptation of what you grow.

pH strips or home liquid test kits give a working range accurate to within about half a point. That is sufficient for most decisions — you do not need to know whether your soil is 5.8 or 6.2 to decide whether to add lime or compost. Run the test in spring and autumn over two or three years to understand whether pH is stable or drifting.

What a Lab Test Tells You

A professional soil test from a cooperative extension service or private lab measures what home kits cannot: precise nutrient levels including calcium, magnesium, sulfur, and trace micronutrients; cation exchange capacity (CEC), which quantifies the soil's ability to hold and release nutrients; organic matter percentage; and in some cases active microbial biomass. The cost is typically $20 to $40 for a comprehensive panel.

Reading the report requires knowing which numbers to prioritize. Organic matter percentage is the most directly compost-relevant figure: below 2% is degraded, 2–4% is working soil, above 4% is well-managed and biologically active. CEC below 10 meq/100g suggests a soil with poor nutrient-holding capacity — one where compost will produce the most dramatic improvement. Any nutrient that falls below the recommended range in the report translates into a concrete amendment recommendation; compost addresses broad deficiencies, but targeted amendments may be needed for specific shortfalls.

How Compost Changes Soil

The mechanism behind most compost benefits is aggregate formation. When organic matter from compost decomposes, bacteria produce sticky polysaccharide compounds that bind soil particles together into stable clusters called aggregates. Inside those clusters, pore spaces form — micro-channels that hold water, allow gas exchange, and provide habitat for soil organisms. The size and stability of aggregates is probably the single most important physical property of garden soil, and compost is the most practical way to build it.

Water-holding capacity is where the improvement is most visible and quantifiable. Research from the USDA shows that each 1% increase in organic matter allows soil to hold an additional 20,000 gallons of water per acre — roughly a 20–30% improvement in water retention across soil types with sustained compost amendment. For a home gardener, this translates directly into fewer waterings during dry spells and plants that survive short droughts without stress.

Cation exchange capacity rises with organic matter content because humic compounds — the stable, long-lived fraction of decomposed organic matter — carry a strong negative charge that holds positively charged nutrient ions (calcium, potassium, magnesium, ammonium) against leaching. Higher CEC means nutrients remain in the root zone rather than washing through with rain. In sandy or degraded soils, building CEC through compost is often more effective than applying more fertilizer.

Compost and Water Management

The water savings from compost-amended soil are worth quantifying specifically because they translate into a real reduction in irrigation costs and labor. Experimental plots comparing compost-amended and unamended soil under equivalent rainfall and irrigation consistently show 20–30% reductions in irrigation water needed to maintain similar soil moisture levels. Over a full growing season in a climate with significant dry periods, this matters both economically and practically.

Erosion reduction is the less-discussed water management benefit. Improved aggregate stability means soil particles resist displacement by rain impact and surface runoff — the physical process that causes topsoil loss during heavy rain events. In sloped beds or gardens that receive intense seasonal rain, the difference in runoff volume between compost-amended and unamended soil can be substantial. The improved infiltration rate — water entering the soil rather than running off the surface — is measurable even after a single season of incorporation.

Biochar and Compost

Biochar is charcoal produced through pyrolysis — heating organic material in a low-oxygen environment — rather than combustion. The resulting material is extraordinarily porous, chemically stable, and resistant to microbial breakdown: biochar added to soil can persist for centuries, unlike most organic matter which cycles within years. The potential for long-term carbon sequestration is what drives interest in it, but the practical garden applications are where the specifics matter.

On its own, fresh biochar is almost inert in soil — the pore structure that makes it useful is empty, and some fresh biochars can temporarily depress plant growth by adsorbing nutrients. Charging it first with compost solves both problems. The method is straightforward: combine biochar with finished compost at a ratio of roughly 1:5 by volume, moisten, and allow it to sit for two to four weeks before application. During that period, microbial communities and nutrients from the compost colonize the biochar's pore structure, priming it for immediate activity when applied to soil.

Charged biochar applied at 5–10% of soil volume by depth improves water retention, provides additional long-term carbon sequestration beyond what compost alone offers, and maintains its structural benefits without reapplication — unlike compost, which is consumed by microbial activity and must be renewed annually. The combination of annual compost inputs and a one-time or periodic biochar amendment covers both short-term biological fertility and long-term physical soil improvement.

Chapter 26

Applying Compost: Methods, Timing, and Quantities

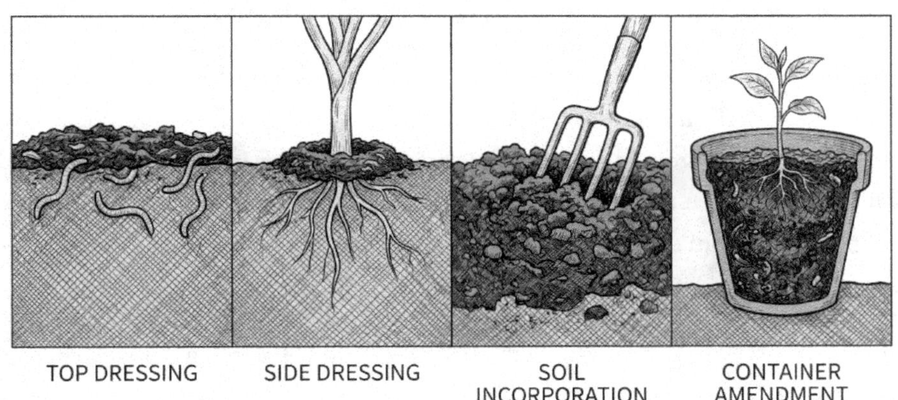

Finished compost sitting in a pile is not doing anything. The biology went quiet during curing, the nutrients are stable but immobile, and the whole point of the last several months — if you hot-composted — or year — if you cold-composted — is not realized until the material enters the soil and the organisms living there get to work. The difference between applying compost correctly and applying it carelessly is, in practice, the difference between a dramatic response and a modest one.

Application method, timing, and quantity all matter independently. Getting one wrong while the others are right blunts the benefit. What follows covers each in enough detail to make informed decisions for your specific soil and planting situation.

The Four Application Methods

Top dressing means spreading one to two inches of compost across the soil surface without working it in. For established perennials, lawns, and trees where disturbing the root zone is undesirable, this is the standard approach: rain and soil organisms draw nutrients downward, earthworms incorporate organic matter, and the compost layer also acts as a mild mulch. It is the gentlest method — and the slowest to produce results, since the compost has to find its own way into the root zone.

Side dressing is a more targeted version of top dressing, applied as a ring around the plant's drip line during the growing season. Heavy-feeding crops — tomatoes, corn, squash, brassicas — respond well to a side dressing of half an inch to an inch applied at mid-season, when they are actively growing and drawing nutrients fast. Keep the compost away from the stem; piled against the crown of a plant creates conditions for rot.

Soil incorporation — working compost into the top six to eight inches before planting — produces the fastest and most complete response because it places the material directly in the zone where roots will develop. Two to four inches of compost tilled into the bed before planting vegetables changes the physical and biological character of the soil more quickly than surface methods. The drawback is that it requires disturbing soil structure, which some gardeners and some planting situations prefer to avoid.

For container plants and potting mixes, compost functions as an amendment within the mix rather than an addition to a larger soil volume. The standard proportion is 20–30% compost by volume of the total mix, combined with the structural components (perlite, bark, coir) that keep containers from compacting. Exceeding 30–40% compost in a closed container increases the risk of anaerobic conditions developing at the base of the pot, particularly if drainage is limited.

Application Timing

Spring applications — incorporated before planting or top-dressed as perennials emerge — capture the full growing season for nutrient uptake. This is the timing most gardeners default to, and for good reason: the soil is warming, microbial activity is accelerating, and plants are about to enter their peak nutrient demand. The limitation is that spring is also the most labor-compressed time in the garden, and compost application competes with everything else.

Autumn applications have a different logic. Post-harvest incorporation refills the organic matter depleted by a full growing season and gives over-winter biology — bacteria, fungi, earthworms active in mild spells — months to begin processing the material before spring. Autumn is also when most gardens generate the bulk of their carbon-rich material: leaf fall, spent plants, straw mulch. Building and incorporating compost in autumn aligns with the seasonal rhythm of most temperate gardens.

At planting time, a modest handful of mature compost worked into each planting hole before setting a transplant or tree gives roots direct access to biologically active material at the moment they begin to establish. Do not overdo this: too much compost concentrated in a planting hole creates a nutrient-rich zone that discourages roots from exploring the surrounding soil. A shovelful per planting hole is appropriate; an entire wheelbarrow load per tree is counterproductive.

How Much to Apply

Vegetable beds used for annual crops benefit from the most generous applications: two to four inches incorporated before each planting cycle, or one to two inches as a spring top dressing in established no-till beds. Heavy-feeding crops grown season after season in the same bed will deplete organic matter faster than plants in rotation, and application rates should reflect that.

Fruit trees and shrubs need less: one to two inches in a two-foot radius from the trunk, pulled back to avoid contact with the bark. The bark contact rule matters because compost piled against wood creates persistently moist conditions at the base of the trunk, inviting crown rot and fungal disease. The application zone should begin several inches from the trunk and extend outward to the drip line.

Lawns respond to a quarter to half inch of screened, finely textured compost broadcast in spring or autumn — a process called topdressing. The particles filter down through the grass to the soil surface over several weeks. More than half an inch smothers grass and takes too long to incorporate. Lawn compost applications are lower-intensity than bed applications but consistently improve turf density and drought resistance when done annually.

Choosing the Right Compost for the Job

Bacterial-dominated compost — the product of hot, actively turned piles with a balanced C:N ratio — is the right material for vegetables and annual flowers. Its high nitrogen content and dense microbial population match the fast growth and high nutrient demand of annual crops. This is also the category that applies most directly to the compost produced through the methods in Book III.

Fungal-dominated compost comes from cooler, less-turned piles processing woody, high-carbon material — wood chips, bark, straw, autumn leaves. Fungi, which tolerate low-nitrogen substrates better than bacteria, dominate the decomposition process. The resulting compost has a more fibrous texture and a C:N ratio tilted toward carbon. For trees, shrubs, and established perennials — which have long-lived root systems and a natural association with mycorrhizal fungi — fungal-dominated compost is a better match than hot-turned bacterial compost.

Vermicompost occupies a different category altogether: lower volume but exceptionally high biological density and immediately plant-available nutrient forms. The castings are most effective when used for their highest-value applications — starting seeds, amending potting mix, side-dressing transplants in their first weeks — rather than as a general soil amendment across large areas where a standard hot or cold compost does the job more economically.

Common Mistakes

Applying immature compost is the most consequential error, and one that is easy to make if the maturity test is skipped. Material that has not completed curing still contains partially decomposed compounds that temporarily immobilize nitrogen as soil bacteria continue processing them — a phenomenon called nitrogen theft. Plants in beds that receive immature compost often display nitrogen deficiency symptoms (yellowing lower leaves, stunted growth) shortly after planting, not despite the amendment but because of it.

Over-application is less common but real: very heavy compost incorporation in a single season — more than four or five inches — concentrates phosphorus and potassium to levels that create nutrient imbalances and in some soils contribute to runoff into waterways. An organic matter percentage above 8–10% in a garden bed is also a point of diminishing returns. Soil does not need to become pure compost; it needs the biological activity that a well-amended but mineral-containing soil supports.

Volcano mulching — piling compost in a mound against the base of a tree trunk — is pervasive in landscaping and consistently damaging. The continuously moist interface between bark and decomposing organic matter accelerates fungal rot in the cambium layer, the living tissue just beneath the bark. The damage is slow and often attributed to other causes, but the mechanism is straightforward. Compost is beneficial to trees in the right location; against the trunk is not that location.

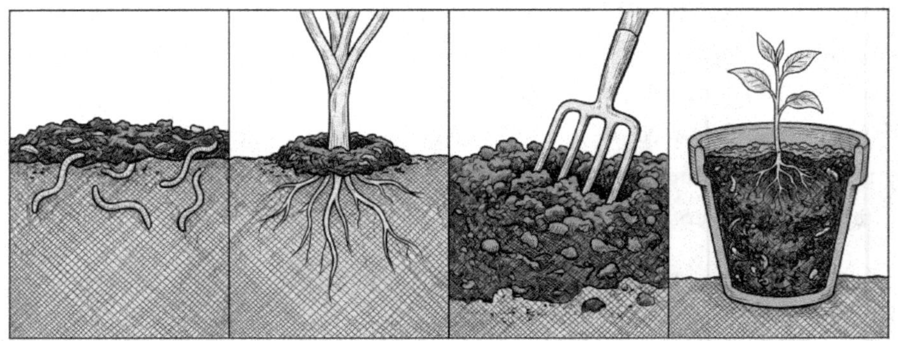

TOP DRESSING
SIDE DRESSING
SOIL
INCORPORATION
CONTAINER
AMENDMENT

Chapter 27

Compost Tea: Brewing Liquid Nutrition for Your Plants

The term "compost tea" covers two entirely different products that share a name and almost nothing else. The version that most gardeners make — steeping compost in water for a day or two without aeration — produces leachate:

a liquid that smells bad, is dominated by anaerobic bacteria, and can be mildly harmful to plants if applied undiluted. The version that works is Aerated Compost Tea (ACT): a 24–36-hour brew with continuous oxygenation that produces a microbially dense, aerobic solution capable of inoculating soil with billions of beneficial organisms per milliliter. Most of the confusion about whether compost tea is effective traces back to this distinction.

ACT is not magic and the claims made for it have ranged from well-supported to extravagant. What the evidence supports is this: a well-made ACT applied as a soil drench increases beneficial microbial populations in the root zone, improves soil biological activity, and supports plant nutrient uptake, particularly in depleted or recovering soils. Foliar application is more variable in results and depends heavily on timing, plant variety, and the specific microbial community in the brew. The science is solid at the soil level; more cautious at the leaf level.

Materials and Equipment

Compost quality is the most important variable and the one most often underestimated. ACT concentrates whatever is living in the source compost — a biologically rich, mature hot compost produces a tea populated by diverse aerobic bacteria and fungi; a low-grade or immature compost produces a tea you would not want near your garden. There is no way to brew a good tea from poor compost. The investment in producing quality compost upstream pays dividends here.

Water requires attention before use. Tap water treated with chlorine or chloramine kills the very organisms you are trying to cultivate. Chlorine off-gasses if water sits in an open container for 24 hours, making dechlorination straightforward for most municipal supplies. Chloramine does not off-gas and requires either a carbon filter, a vitamin C tablet (ascorbic acid — one gram per 10 gallons), or an alternative water source. Well water is usually suitable without treatment.

The brewing setup is simple: a five-gallon bucket, an aquarium air pump capable of producing 0.05 cubic feet per minute or more per gallon of water, flexible airline tubing, and two or three aeration stones positioned at the base of the bucket. The goal is vigorous, continuous agitation throughout the entire brew — a steady stream of fine bubbles from the bottom up. This keeps dissolved oxygen high enough to maintain aerobic conditions and prevents sediment from settling and going anaerobic.

Unsulfured molasses — one to two tablespoons per five gallons — serves as a carbohydrate food source that allows fast-reproducing bacteria to expand their populations during the brew. Kelp meal (one tablespoon per five gallons) adds trace minerals and growth compounds that favor fungal development, producing a more balanced microbial community. Both are optional but produce measurably more active teas than compost and water alone.

Step-by-Step Brewing

Start with roughly one cup of finished compost per gallon of water, placed in a mesh bag or fine-weave fabric that allows water contact while keeping particles contained. Fill the bucket with dechlorinated water at 65–75°F (18–24°C) — temperature is more important than most instructions suggest. Below 60°F, bacterial reproduction slows significantly; above 80°F, the wrong organisms proliferate. Add molasses and any optional additives.

Run the pump continuously for 24 to 36 hours. Check the tea at 18 hours: it should smell earthy, slightly sweet, and alive — a smell associated with healthy garden soil after rain. A sour or sulfurous smell indicates that anaerobic conditions have developed somewhere in the bucket, most likely around a settled sediment zone. This batch should be diluted heavily and used as a non-critical soil drench or discarded.

Strain the finished tea through cheesecloth or a fine-mesh strainer and use it within four hours. This is the constraint that limits the practicality of compost tea for large-scale application: the aerobic bacteria reproduce rapidly during brewing but die off equally rapidly once the oxygen supply stops. Finished ACT left sitting for eight hours is a different, less useful product than the one you brewed. Schedule brewing to coincide with application.

Application Methods

Soil drenching — watering the root zone directly — is where ACT produces the most reliable results. Apply at a rate of about one gallon per 10 square feet, in the early morning or evening when UV radiation is lowest. Dilute the tea one-to-one with dechlorinated water if applying frequently; full-strength tea applied weekly is more intensive than most soils need.

Foliar spraying introduces the microbial community to the leaf surface, where some organisms compete with fungal pathogens and others produce compounds that stimulate plant immune responses. The window for effective foliar application is narrow: early morning, when stomata are open and leaf surface temperature is low. A fine mist applied to both leaf surfaces dries within an hour or two, which is enough contact time for colonization. Avoid spraying in direct sun or when temperatures will exceed 85°F the same day.

Troubleshooting and Realistic Expectations

Foul-smelling tea is the most common problem and almost always indicates anaerobic conditions: pump failure during the brew, insufficient aeration for the volume of water, or a compost source that was not fully mature and carried anaerobic pockets into the bucket. A tea that smells like sulfur or sewage should not be applied to plants. Heavily diluted, it can go on a compost pile or a fallow area of the garden; otherwise, discard it.

Excessive foam usually means too much molasses. At one to two tablespoons per five gallons this is unlikely, but some compost sources are sugar-rich enough that even small molasses additions push into foam territory. Reduce or eliminate the molasses and run the batch again.

Compost tea is not a substitute for compost, and it is not a substitute for fertilizer when a specific nutrient deficiency needs correction. Its value is biological — introducing and expanding the microbial community that makes nutrients accessible and soil structure resilient. Gardeners who see the most consistent results with ACT are those using it to maintain already biologically active soils, or to accelerate recovery in soils that have been depleted, tilled heavily, or treated with synthetic inputs. As a standalone intervention in severely deficient soil, it underperforms without the foundational compost amendment that healthy soil biology requires.

Beyond Compost: Mulching, Cover Crops, and Crop Rotation

A garden bed managed on compost alone is like an engine running on one cylinder. The organic matter improves soil, but the improvement is gradual, and the inputs required to maintain it are continuous. Mulching, cover crops, and crop rotation are the other cylinders: they reduce the demand that compost has to meet, build fertility through different mechanisms, extend the effects of each compost application, and collectively produce soil health that compounds over years rather than resetting each season.

None of these practices require equipment, expertise, or significant time investment beyond what the garden already demands. What they require is understanding what each one does and when it is the right tool.

Mulching

Organic mulch — straw, wood chips, shredded leaves, grass clippings — laid two to four inches deep over the soil surface does four things simultaneously: it holds moisture by reducing evaporation, moderates soil temperature, suppresses weed germination by blocking light, and eventually breaks down into organic matter that feeds soil biology. The first three benefits appear within the first week of application. The fourth accumulates over months.

Straw mulch is the standard for vegetable beds: it breaks down within a single growing season, is light enough to push aside for planting, and does not harbor pathogens the way hay (which contains seed heads) does. Wood chips are better suited to perennial beds and pathways where slower breakdown is acceptable and deeper structure is needed. Shredded leaves — run through a mower or leaf shredder to prevent matting — are among the most effective and freely available mulch materials for most temperate gardens. Grass clippings work well applied thinly and mixed with other materials; a thick layer of fresh clippings mats down and goes anaerobic.

The single most important rule of mulch application is the gap at the stem. Two to three inches of space between the mulch layer and the base of any plant prevents the moisture retention that mulch is designed to provide from becoming a rot condition at the crown. Apply generously across the open bed surface; pull back firmly around every stem and trunk.

Decomposed mulch at the end of the season does not need to be removed. Incorporate it lightly into the soil surface before the next planting season, or simply plant through it and let the ongoing decomposition contribute organic matter. Each season's mulch becomes part of the following season's soil organic matter — a closed loop that reduces the quantity of external compost needed to maintain fertility.

Cover Crops

A cover crop is a plant grown not for harvest but for what it contributes to the soil when it is terminated and incorporated. The category covers a wide range of species with different mechanisms, and choosing among them is a matter of matching the mechanism to the need.

Nitrogen-fixing legumes — clovers, field peas, hairy vetch, fava beans — form root associations with Rhizobium bacteria that capture atmospheric nitrogen and store it in root nodules. When the crop is terminated and incorporated, that nitrogen enters the soil organic matter pool and becomes available to following crops. A well-established stand of hairy vetch can contribute 100 to 200 pounds of nitrogen per acre over a season — the equivalent of several substantial compost applications in a single biological cycle. For the home gardener managing a few hundred square feet, the scale is smaller but the mechanism is identical.

Biomass builders like buckwheat, oats, and winter rye are chosen for fast above-ground growth rather than nitrogen fixation. Their value is the large quantity of organic matter they generate quickly — buckwheat can reach two feet in height in five weeks — and their root systems, which penetrate compacted layers and leave channels in the soil when they decompose. Winter rye planted in autumn and terminated in spring is one of the most reliable cover crops for temperate gardens, suppressing weeds through winter and providing a substantial biomass to incorporate before the growing season.

Dense-growing species like buckwheat and crimson clover also function as living weed suppressors, shading the soil surface effectively enough to prevent most weed seed germination during the period they are established. This matters most in the period between crops — the gap weeks when a bed is empty and weed pressure is highest. A quickly established cover crop closes that window.

Termination method depends on the species and your tillage approach. Annual cover crops can be mowed and left on the surface as a mulch layer, or cut and incorporated with a fork or tiller. Perennial clovers are cut repeatedly and left to regrow if used as a living mulch between rows. The key rule is to terminate before seed set — an unharvested cover crop that goes to seed becomes the next season's weed problem.

Crop Rotation

Rotating plant families through different beds each season breaks the cycles of pests and diseases that accumulate when the same crops return to the same ground repeatedly. Clubroot in brassicas, tomato blight, and root-knot nematodes all depend on building populations in soil through repeated host-plant presence. Remove the host for one or two years and populations decline to manageable levels without intervention.

For home vegetable gardens, a workable rotation groups crops into four botanical families: nightshades (tomatoes, peppers, eggplant, potatoes), brassicas (cabbage, broccoli, kale, turnips), legumes (beans, peas), and roots and alliums (carrots, onions, garlic, beetroot). Each family moves to a new bed each season, completing a four-year cycle before returning. The rotation does not need to be mathematically perfect — approximate rotation reduces pest buildup significantly even when beds don't line up precisely.

The intersection of rotation and compost application is where planning pays off. Brassicas and nightshades are heavy feeders and benefit from compost incorporation before planting. Legumes fix their own nitrogen and should receive compost only if the soil is genuinely depleted — over-fertilizing before legumes encourages leafy growth at the expense of root nodule formation. Roots and alliums perform better in soil that was composted the previous season rather than immediately before planting; fresh, rich compost encourages forking in carrots and excessive leafy growth in onions.

A simple garden map — sketched on paper or maintained in a notebook — prevents the memory failures that make rotation plans collapse. Record what was planted where each season, note which beds received compost, and mark any pest or disease issues that arose. Over three or four years, patterns emerge that make subsequent planting decisions significantly more informed.

Companion Planting

Companion planting — growing specific plants together to mutual benefit — has a long practical history and a shorter scientific one. The mechanisms that are well-supported include: aromatic herbs and flowers that deter or confuse pest insects through scent; flowering plants that attract and sustain beneficial insect predators; and dense ground-cover plants that suppress weeds in the spaces between taller crops.

The classic three sisters combination — corn, climbing beans, and squash grown together — is functional rather than folkloric. Corn provides a climbing structure for the beans; beans fix nitrogen that feeds the corn and squash; squash shades the soil with its large leaves, reducing moisture loss and suppressing weeds. Each plant supports the others structurally and nutritionally. Transplanted to a home garden scale, the combination works in beds as small as four by four feet.

Companion planting complements composting by addressing the pest and disease pressures that would otherwise require intervention — chemical or physical — that costs time and, in some cases, damages the soil biology that compost builds. A garden that uses both is not working twice as hard; it is working the same amount in a more integrated way.

PART VI

COMPOSTING AND THE
WORLD SUSTAINABILITY

Beyond the Backyard

Chapter 29

Composting and the Climate: Numbers, Facts, and Real Impact

Food waste is responsible for approximately 8% of global greenhouse gas emissions. That figure comes from the Food and Agriculture Organization of the United Nations and includes the emissions generated across the entire food production chain for food that is never eaten — growing, transporting, refrigerating, and then landfilling it. Eight percent is roughly the combined aviation and shipping industries. It is a number large enough that what happens at the end of the food chain matters at the scale of climate.

Composting does not fix this problem. What it does is change the chemistry of what happens to organic matter that reaches the end of its useful life — and that chemical difference, multiplied across millions of households and commercial operations, is meaningful in a way that is worth understanding accurately rather than either overstating or dismissing.

Organic Waste and Methane

The difference between composting and landfilling comes down to oxygen. In an active compost pile, organic matter decomposes aerobically — bacteria using oxygen to break down carbon compounds, releasing carbon dioxide (CO_2) as the main carbon byproduct. In a landfill, organic matter buried under other waste decomposes anaerobically — without oxygen — producing methane (CH_4) as the primary carbon byproduct. Over a 20-year time horizon, methane's global warming potential is 80 times that of carbon dioxide. Over 100 years, the ratio falls to around 28, but the near-term impact is what matters for the climate trajectories currently in play.

Municipal landfills capture some of this methane for energy production, but capture rates are incomplete — typically 50 to 90% of generated methane, depending on landfill design and management. The rest escapes. Even in the best-managed landfills, significant methane emissions occur during the years before a cell is capped and gas collection infrastructure is installed. Diverting organic waste to composting removes the substrate for anaerobic decomposition before it reaches the landfill.

Carbon Sequestration

The carbon captured in compost-amended soil is not all equivalent. Some fraction of compost-derived carbon is rapidly consumed by soil organisms and released as CO_2 within the first growing season. Another fraction — the humic compounds formed during decomposition — is chemically stable and persists in soil for decades or centuries. This stable fraction is what constitutes genuine carbon sequestration, as opposed to temporary carbon storage.

Estimates of sequestration potential vary with soil type, management practices, and the specific fraction of humus formed, but the direction is consistently positive. Research from the Rodale Institute and others has documented that converting conventionally farmed cropland to organic management with regular compost inputs can shift those soils from a net carbon source to a net carbon sink over a period of several years. The global potential, if 10% of agricultural soils received annual compost applications, is estimated at the equivalent of several gigatons of CO_2 per year — a contribution meaningful at the scale of national emissions targets.

Compost vs. Synthetic Fertilizer

The Haber-Bosch process — the industrial synthesis of ammonia for synthetic nitrogen fertilizer — consumes approximately 1% of total global energy use and accounts for roughly 1.5% of global CO_2 emissions annually. That is the upstream cost of the nitrogen contained in synthetic fertilizers, paid before the product reaches any field. Compost generates nitrogen through biological cycling that uses solar energy captured by plants, requiring no fossil fuel inputs beyond the minimal energy of transport and turning equipment.

Once applied, synthetic nitrogen fertilizer generates a secondary emission problem through nitrous oxide (N_2O). When mineral nitrogen is applied to soil in excess of what plants can uptake immediately, soil bacteria convert it through a process called denitrification into N_2O — a greenhouse gas with 273 times the warming potential of CO_2 over 100 years. Compost releases nitrogen slowly and in biologically mediated forms that plants take up progressively, reducing the nitrogen surplus that drives N_2O production. This does not mean compost replaces synthetic fertilizer in all situations, but the full lifecycle comparison is considerably less favorable to synthetic inputs than the point-of-application nitrogen numbers suggest.

Individual Impact: Honest Numbers

The average US household generates approximately 650 pounds of compostable waste per year — food scraps, yard waste, and compostable paper products. If composted rather than landfilled, the rough CO_2-equivalent reduction per household is estimated at 0.5 to 1.0 metric tons per year, depending on the fraction of organic material involved and regional landfill practices. That is comparable to eliminating several hundred miles of annual car travel.

The honest framing requires acknowledging scale. Individual household composting, even if universal, would not by itself close the gap on the food system's climate impact — the upstream emissions from food production dwarf the end-of-life processing emissions. Systemic interventions — changes in agricultural practice, supply chain efficiency, food waste reduction at the production and retail level — move larger numbers. What composting offers at the household level is a genuine and measurable contribution combined with the soil benefits that are the primary reason to do it. The climate case is secondary and real; it should not be inflated into the central argument.

The Soil Microbiome and Human Health

The connection between soil biodiversity and human health is an emerging research area that had not reached scientific consensus at the time this edition was written, but the evidence accumulating is worth understanding in broad terms. Conventional agricultural soils treated repeatedly with synthetic fertilizers, herbicides, and fungicides show consistently reduced microbial species richness compared to organically managed soils — fewer genera, smaller populations, and lower functional diversity.

The gut-soil axis hypothesis proposes that the microbiome of the plants we eat — colonized by soil organisms and affected by soil microbial diversity — influences the diversity and composition of our gut microbiome. The supporting evidence includes studies linking organic produce consumption to higher gut bacterial diversity in consumers, and mechanistic research on plant compounds whose production is mediated by specific soil fungal and bacterial associations. The causality is not fully established, but the correlation is consistent enough to take seriously.

For the home gardener growing in compost-amended soil, the practical implication is that the quality argument for homegrown food is broader than freshness and the absence of pesticide residue. A vegetable grown in biologically active soil may carry a more diverse microbial community than one grown in sterile substrate under controlled conditions. Whether and how much this matters for human health remains an open question — but it is a more interesting question than it was a decade ago.

Chapter 30

The Economics of Composting: What You Save, What You Create

Composting costs time and nothing else, if you already have organic waste. The inputs are material that would otherwise be discarded; the process requires labor measurable in minutes per week; and the output replaces purchased products with documented market prices. The economics of home composting are not complicated, but they are consistently underestimated by the people who do not compost and occasionally overstated by the people who do.

What You Save at Home

Organic fertilizer prices have risen significantly over the past decade. At current retail, bagged finished compost runs $8 to $15 for 40 pounds; vermicast sells for $30 to $50 per 20 pounds; commercial compost tea products reach $25 to $40 per quart. A home composter producing 200 to 400 pounds of finished compost per year — which a moderately managed household pile does routinely — saves $40 to $150 in purchased soil amendments at minimum, and considerably more if vermicast or specialty compost would be the comparison.

Water savings are less often calculated but add up over a growing season. Compost-amended beds requiring 20–30% less irrigation than unamended ground — a figure supported by multiple experimental comparisons — translates to real reductions in water bills in households on metered supply and real reductions in watering labor in all households. In drought-prone regions or climates with dry summers, the water retention improvement is among the most practically valuable outcomes of sustained compost application.

In pay-as-you-throw waste disposal programs — now active in hundreds of US municipalities and expanding — residents pay per unit volume of landfill-bound waste. Diverting food scraps and yard waste to home composting directly reduces the chargeable volume. Households that compost consistently report reductions of 30–50% in their landfill-bound waste stream, which translates to lower monthly waste disposal costs proportionally.

Municipal and Societal Value

The cost of managing organic waste through conventional landfilling — tipping fees, transportation, leachate management, gas capture infrastructure, and long-term capping — runs between $30 and $80 per ton in the United States, depending on region and landfill design. Composting programs process the same material at $20 to $50 per ton and produce a marketable product at the end. The economics favor composting at the systems level even before accounting for the avoided methane emissions that create additional cost liabilities under emerging carbon pricing frameworks.

Wholesale finished compost sells at $15 to $40 per ton in bulk markets; retail bagged compost at $200 to $400 per ton equivalent. Municipal programs that produce and sell compost can offset a substantial portion of program operating costs through product revenue. Several large urban programs — including composting initiatives in San Francisco, Austin, and Seattle — have demonstrated financial self-sufficiency or near-self-sufficiency once at operating scale.

Urban Farming and Local Food Systems

Degraded urban soil — often chemically contaminated, compacted, or so low in organic matter that it barely qualifies as soil — is the primary limiting factor for urban food production in most cities. Compost is the only amendment that addresses all the relevant deficiencies simultaneously: organic matter, structure, biology, and nutrient content. Vacant lots transformed into productive growing space through sustained compost inputs exist in every major US city, and the pattern is consistent enough to be treated as a reliable intervention rather than an exceptional outcome.

The economic case for urban farming includes reduced supply chain costs, fresher produce at lower distribution cost, and — in communities with limited grocery access — improved food security. Compost is not sufficient to make urban farming economically viable on its own, but it is necessary: urban farming without compost input runs against the soil conditions that most urban sites start with.

Worm Farming as a Small Business

The market for worm castings is real and growing. Premium retail vermicast at $3 to $5 per pound, and wholesale at $0.50 to $1.50 per pound, makes small-scale worm farming potentially profitable for operators willing to manage the volume and sales logistics. A system processing 200 pounds of food waste per week — achievable in a space of 200 to 300 square feet with stacked flow-through vermicomposting units — produces roughly 50 to 80 pounds of castings per week, generating $25 to $120 per week at wholesale pricing.

The practical challenges are less about the worms and more about distribution: finding consistent buyers at prices above the breakeven cost of feed material acquisition, labor, and system maintenance. Farmers markets, local garden centers, CSA operations, and community garden programs are the most reliable local sales channels. Worm tea marketed as a liquid soil amendment adds a second revenue stream with low incremental cost, though compliance with state fertilizer labeling regulations applies in most US states and should be researched before marketing.

Community Composting Economics

Neighborhood-scale composting programs — drop-off sites, shared equipment, organized collection from participating households — occupy the space between individual home composting and full municipal programs. Their economics depend heavily on volunteer structure and grant funding in early years, with a path toward sustainability through compost sales, small service fees from participating households, and in some cases municipal contracts for food waste processing.

Programs that have reached sustainable operation share several characteristics: consistent feedstock supply (usually 20 or more regularly participating households), a designated output market (community garden beds, urban farms, or school grounds that absorb the finished compost), and a volunteer-or-staff management structure with clear continuity beyond founding members. The organizations that fail tend to do so at the transition point when founding energy fades before institutional structures take over. Building program infrastructure — documented processes, trained volunteers, governance — before that transition point is the most important design decision.

Composting in Schools and Communities: Building the Next Generation

On a Tuesday morning in early May, a class of nine-year-olds in Sacramento is sorting lunch waste into three bins: food scraps, liquids, and everything else. The food scraps go to a worm bin maintained by a rotating student committee. The worm bin feeds a raised bed where the same class grew the salad greens that appeared in the school cafeteria two weeks ago. The children have not been told that what they are doing is environmentally significant. They have been told that the worms need feeding, the plants need compost, and the lettuce needs harvesting. The rest follows.

School composting programs succeed when they are embedded in something students care about rather than presented as an obligation. The worm bin that connects to the school garden that connects to the cafeteria salad bar is a feedback loop that children understand intuitively. The composting is the mechanism, not the point.

The Educational Case

Composting crosses more curriculum areas than almost any other hands-on school activity. Biology: the life cycles of decomposers, the microbial community, the chemistry of decomposition. Math: weight measurement, ratio calculations, temperature monitoring, graphing pile temperatures over time. Environmental science: the carbon cycle, waste reduction, soil ecology. Civic responsibility: shared maintenance of a communal resource, decisions about what inputs the program accepts, outreach to other classes and families. A well-designed school composting program is not an add-on to the curriculum; it is a cross-disciplinary practicum.

Research on outdoor and hands-on education consistently shows improved retention, increased engagement, and more durable attitude formation compared to classroom-only instruction. Students who have physically turned a compost pile, observed worm castings under a magnifying glass, and eaten food grown in compost they helped produce have a different relationship to the subject than students who read about it. The behavioral change evidence is also positive: children who participate in school composting programs are more likely to compost at home and more likely to introduce composting to their families.

Setting Up a School Program

Stakeholder alignment precedes everything else and is where most well-intentioned school composting initiatives stall. The principal needs to understand that the program will require some administrative overhead and will produce visible results worth the investment. Custodial staff need to be involved from the start, not informed after the fact — they manage the physical space, handle the waste infrastructure, and can make or break a program's day-to-day operations. Cafeteria staff need a sorting protocol that does not significantly complicate their service flow. Parent volunteers provide energy for setup and early maintenance; the program needs to reach sustainability without depending on any specific volunteer indefinitely.

System choice depends on school type and available space. A school with outdoor grounds can run an outdoor bin or hot pile that handles high volumes, involves students in turning and monitoring, and produces compost for a school garden. A school without outdoor access can run one or more worm bins that fit in a classroom or hallway — lower volume but complete in themselves and directly connected to student oversight. Bokashi systems work in high-density urban schools where outdoor access is impossible and worm management is a constraint, though the two-step process requires a community garden or partner site for the second stage.

Curriculum integration does not require redesigning existing lesson plans. A composting temperature graph covers data collection and graphing skills. A worm bin population estimate covers estimation, sampling, and multiplication. A cafeteria waste audit covers measurement, percentages, and data communication. The composting program provides the data; the academic objectives determine how it is used. Most teachers find that composting activities slot into existing units more easily than expected once the connections are made explicit.

Successful Programs

The Los Angeles Unified School District composting program, operating across hundreds of schools in one of the largest school districts in the United States, demonstrates that large-scale cafeteria composting is logistically achievable without specialized infrastructure. The key lessons from LAUSD: sorting education for students is fast to implement and dramatically reduces contamination; dedicated waste station placement during service reduces the time burden on cafeteria staff;

and visible, accessible composting destinations (on-site bins or regular collection) maintain participation better than abstract disposal.

Farm to School composting initiatives connect cafeteria waste directly to the soil that grows school garden produce. Where a school has both a cafeteria food waste stream and a garden, the composting program becomes a physical demonstration of the closed loop that most environmental education only describes. Programs in Vermont, Oregon, and Minnesota have documented both educational outcomes and measurable waste reduction, providing replicable models for districts considering similar programs.

Advocacy: Spreading the Practice

The most effective entry point for converting skeptics to composting — whether in a school, a neighborhood, or a workplace — is economics rather than ecology. The ecological arguments are accurate but easily dismissed by someone who does not yet have a stake in the outcome. The economic arguments — lower waste costs, lower fertilizer costs, better yields — are concrete and immediately verifiable. Lead with what composting saves, not with what it saves the planet, and the audience that is unreachable on environmental grounds becomes reachable on practical ones.

Grassroots composting advocacy has produced some of the most innovative program models operating today. Compost Queens in San Antonio built a neighborhood pick-up network that now serves thousands of households through a subscription model, turning food scraps into compost sold back to participating households. Incredible Edible in Todmorden, England, demonstrated that growing food in public spaces — including food grown in community-composted soil — creates community cohesion effects that go beyond gardening. 3000 acres in Melbourne has mapped and activated unused growing spaces across the city using compost as a central soil-building tool. Each of these started with a small group of people who did not wait for institutional permission.

Chapter 32

The Future of Composting: Technology, Policy, and What's Coming

T hirty years ago, curbside recycling was a niche municipal service in a handful of progressive cities. Today it is infrastructure — unremarkable, universal, and legally mandated in most jurisdictions. Composting is at roughly the stage recycling was in the early 1990s: growing, inconsistently available, dependent on early adopters, and positioned at the beginning of a policy and technology curve that will look, in retrospect, like the obvious direction. The question is not whether composting will become standard infrastructure; it is how fast and in what form.

Smart Technology for Home Composting

IoT-enabled compost monitors have moved from prototype to commercially available products in the last several years. Devices from companies including Edyn, Harvest Hero, and smaller developers now provide real-time temperature, moisture, and CO_2 readings from sensors embedded in compost piles, connected to smartphone apps that alert when turning is needed or conditions drift outside the optimal range. For composters managing multiple piles or hot composting on a precision schedule, the diagnostic value is genuine. For the casual cold composter, the sensors are unnecessary — the pile will tell you what it needs through its own indicators.

AI-assisted identification apps can now accurately classify most common compostable materials from a smartphone photo, provide C:N estimates, and flag inputs that should not enter a home compost pile. Several garden apps have integrated composting troubleshooting — step-through diagnostic tools that work through symptoms (smell, appearance, temperature, moisture) to identify likely problems and recommend specific interventions. The quality of these tools has improved enough that they are genuinely useful for beginners navigating their first pile failures.

Electric countertop composters — Lomi, Vitamix FoodCycler, Reencle, and a growing field of competitors — occupy a specific market: households willing to pay $300 to $700 for a kitchen appliance that eliminates the outdoor composting step entirely. The honest assessment of current-generation machines is mixed. Most primarily dehydrate and grind food waste rather than composting it in the biological sense; the output requires further soil contact to complete decomposition and cannot be used at full strength as a plant amendment. The Reencle, which uses a microbial culture to begin active decomposition, produces an output closer to genuine pre-compost. For households with no outdoor access and no appetite for live biology management, these machines provide a viable diversion option; they are not a replacement for compost in the soil amendment sense.

Industrial and Large-Scale Innovation

In-vessel composting — processing organic waste in enclosed, mechanically aerated chambers with controlled atmosphere — has matured into reliable municipal-scale technology. Systems from manufacturers including Engineered Compost Systems and BDP Industries process hundreds of tons per day with odor containment, controlled retention times, and consistent output quality. The technology is not new, but its deployment has accelerated as tipping fees rise and organic waste diversion mandates create economic pressure to move material out of landfills.

Anaerobic digestion at scale captures the methane that composting intentionally avoids producing. Large biodigester facilities process organic waste in sealed, oxygen-free vessels, capturing the methane as biogas for electricity generation or vehicle fuel, and producing a nutrient-rich digestate as a residual soil amendment. The energy recovery is the key advantage over aerobic composting for very high-volume organic waste streams. Several European countries have built anaerobic digestion into their municipal waste infrastructure as the primary large-volume organic waste pathway, with composting as the secondary system for material streams that do not generate sufficient biogas to justify the capital cost.

Small-scale biodigesters — household and farm-scale anaerobic systems that produce cooking gas from food scraps and manure — are well-established in India, Nepal, China, and other parts of Asia and Africa, where millions of installations have been operating for decades. Entry into Western markets has been slow, limited by regulatory complexity, higher labor costs, and competition from grid-connected renewable energy. Several manufacturers are now producing pre-engineered household biodigester kits for the US and European markets, but uptake remains limited outside early adopters.

Policy and Legislation

California's SB 1383, signed in 2016 and phased in through 2022, is the most comprehensive organic waste diversion mandate in the United States. It requires 75% reduction in organic waste sent to landfills by 2025, mandates that a significant portion of recovered material be used in food-insecure communities, and creates a regulatory framework that treats organic waste diversion as an infrastructure obligation rather than a voluntary practice. Several other states — Vermont, Massachusetts, New York, Connecticut — have enacted similar mandates at different scales. The trend is clear and irreversible.

Extended Producer Responsibility (EPR) frameworks — already governing electronics, paint, and packaging waste in multiple US states — are beginning to reach food packaging and food service materials. Under EPR, producers bear financial responsibility for end-of-life management of their products, creating economic incentives to design compostable packaging that actually ends up in compost streams rather than landfills. The policy is early-stage in the organic waste context, but the direction is established.

Carbon markets are beginning to price soil carbon sequestration in ways that could eventually create financial incentives for composting at the farm scale. Protocols for soil organic carbon credits already exist under several voluntary carbon market standards. The monitoring and verification requirements currently make small-farm participation expensive relative to the credit value, but the infrastructure is being built, and several agricultural states are exploring whether to subsidize measurement costs to encourage participation.

Where Composting Will Be in Twenty Years

The most probable twenty-year trajectory places composting where recycling is today: a standard municipal service available to most residents in developed countries, legally mandated for commercial food generators, and increasingly integrated with anaerobic digestion systems that handle the highest-volume organic waste streams while composting handles the material that benefits most from biological treatment rather than energy extraction. Home composting will remain a distinct and valuable practice for gardeners regardless of what the municipal system provides, for the same reason that people still grow tomatoes even though supermarkets exist.

The longer-term vision — healthy soil recognized as a form of public infrastructure, with composting as one of its primary maintenance inputs — is not fantastical. The science connecting soil health to food quality, water management, climate stability, and biodiversity is converging. The policy frameworks are being built. The technology is available and improving. The gap between where composting is now and where it could be is primarily an information and incentive gap, not a technical one. That is the kind of gap that closes within a generation when the conditions are right.

Frequently Asked Questions: 30 Answers for Every Stage

Every composting question worth asking has been asked many times. What follows are the thirty that appear most consistently — from beginners who haven't started yet, from experienced composters who've hit a specific problem, and from urban composters working within constraints that the standard answers don't address. The answers here are direct. Where the honest answer is "it depends," the dependency is named specifically rather than left to the reader to resolve.

Getting Started

Q1: Can I compost in an apartment?

Yes, and with more options than most people realize. A worm bin handles vegetable and fruit-based kitchen waste and fits under a sink or in a closet. Bokashi handles the complete range of food waste including meat and dairy in two sealed buckets. Electric countertop composters require only counter space and no biological management. The constraint in apartments is generally what to do with the output, not the composting process itself — which is where community garden drop-offs, building composting programs, and municipal organics collection become relevant.

Q2: How much time does it actually take?

A worm bin requires five to ten minutes per week once established. A cold pile can run on as little as thirty minutes per month. Hot composting during the active phase requires thirty to forty-five minutes per turning, every three to five days — roughly two to three hours per week for four to eight weeks per batch. The ongoing maintenance time of any system is much lower than the setup and learning curve time, which is front-loaded.

Q3: What's the simplest way to start today?

If you have outdoor space: designate a pile area, start adding kitchen scraps and yard waste in a rough 1:1 ratio by volume, water if it gets dry, and wait. If you have no outdoor space: order red wigglers online, find or buy a container with drainage, add damp shredded cardboard bedding, introduce worms, and begin feeding the next day. Neither requires purchasing anything expensive or following a complicated protocol.

Q4: Do I need special equipment?

No. Open-pile composting requires nothing except a designated area. Contained composting can use purpose-built bins, but a wooden pallet enclosure or a cylinder of wire mesh functions identically. Vermicomposting needs only a container with drainage and aeration — a plastic storage tub with holes drilled in the lid and base works as well as a $200 stacking vermicomposter for small volumes.

Q5: Can I compost if I only generate a small amount of waste?

Cold composting and vermicomposting are specifically suited to small, intermittent waste streams. A hot pile requires enough material to build the minimum mass in a single session, but a cold pile or worm bin accepts additions of any size on any schedule. The output timeline extends proportionally, but the process works.

Materials and Inputs

Q6: What can and can't I compost?

Everything plant-based can go into a standard aerobic pile: vegetable and fruit scraps, coffee grounds, tea leaves, paper, cardboard, grass clippings, leaves, garden trimmings, wood chips, eggshells. Avoid in a standard pile: meat, fish, dairy, cooked oily food, pet waste, diseased plant material, persistent weed seeds. The full materials guide with C:N ratios is in Book II Chapter 2.

Q7: Can I compost meat and dairy?

In a properly managed hot pile that reaches and sustains 140°F (60°C), yes — the heat kills pathogens and the thermophilic decomposition is fast enough that odor is minimal if the material is buried in the center. The more accessible method for most households is bokashi, which handles meat, dairy, fish, and cooked foods without odor issues and without requiring the management precision of an active hot pile.

Q8: Can I compost pet waste?

Dog and cat waste should not go into compost intended for food gardens, due to the pathogen risk from organisms like Toxoplasma and Campylobacter that are not reliably killed at standard home composting temperatures. Dedicated pet waste composters — separate systems run at higher temperatures or processed through anaerobic methods — can safely handle this material for use in ornamental beds only. Herbivore manure (rabbits, guinea pigs, hamsters) is safe in standard compost and a valuable nitrogen source.

Q9: What's the ideal green-to-brown ratio?

By volume, roughly 1:1 for a pile you want to heat actively; leaning toward more browns if odor is a concern; leaning toward more greens if the pile is slow to heat and materials are dry. The 25–30:1 C:N ratio target is the underlying science; the 1:1 volume rule is a practical approximation that works for most combinations of common materials.

Q10: Can I use worm bin castings directly in the garden?

Yes, without restriction. Vermicompost is the most stable, biologically active, and immediately plant-available compost product in this book. It can be applied to seedling trays, mixed into potting soil, side-dressed around established plants, or top-dressed over lawn surfaces. The only limitation is quantity — home vermicomposting systems produce a relatively small volume, which is why the castings are best reserved for high-value applications rather than broad-area amendment.

Troubleshooting

Q11: My pile smells bad. What's wrong?

The smell identifies the problem specifically. Ammonia smell: too much nitrogen, not enough carbon — add browns. Rotten-egg or sulfur smell: anaerobic conditions — turn the pile, check moisture (too wet), reduce dense or waterlogged material. Sweet, vinegary smell: partially anaerobic fermentation, often from fruit scraps — turn and add carbon-rich material. A well-functioning pile smells like earth after rain.

Q12: My pile isn't decomposing.

Check four variables in this order: moisture (the pile should feel like a wrung-out sponge — if it's dry, decomposition stalls completely), C:N ratio (add greens if you have mostly browns), mass (a pile under one cubic meter won't heat up; cold composting will still work but slowly), and oxygen (a compacted pile needs turning). Address the most obvious deficiency first.

Q13: I have pests in my compost.

Rodents are almost always a response to food scraps accessible at the pile surface or buried shallowly. Bury scraps deeply in the center, cover the pile with a welded wire bottom, and avoid meat, dairy, and cooked food in open piles. Flies and insects are normal decomposers — only a problem if the pile is producing excessive liquid or has surface-exposed food. Raccoons and larger animals are deterred by enclosed bins or buried wire mesh around the base.

Q14: My pile is too wet or too dry.

For a wet pile: add dry carbon materials (wood chips, cardboard, straw), turn to expose wet interior to air, cover in rain. For a dry pile: add water slowly while turning, targeting even moisture throughout rather than wetting the surface only, add nitrogen-rich green material. Both conditions resolve within a few days of correct intervention.

Q15: White fuzzy growth appeared on my pile.

Almost certainly beneficial fungal mycelium — Actinomycetes or fungal hyphae that are normal and healthy parts of the decomposition community. White and threadlike is good. Blue, green, or black growth on bokashi output indicates contamination from unwanted mold species, which usually means air entered the sealed container. In an open compost pile, almost all visible fungal growth is beneficial.

Technique-Specific

Q16: What's the difference between hot and cold composting?

Speed, management requirement, and output quality. Hot composting produces finished compost in four to eight weeks, kills weed seeds and pathogens, requires active management with turning and monitoring. Cold composting produces finished compost in six months to two years, cannot reliably kill weed seeds or pathogens, and requires minimal management. Both produce nutritionally valuable compost; the choice depends on how much material you have, how fast you need output, and how much time you want to invest.

Q17: How is vermicomposting different from regular composting?

Different organisms, different inputs, different output. Vermicomposting relies on worms and the microbial community associated with them; standard composting relies primarily on bacteria and fungi. Vermicomposting handles small, ongoing food waste additions; hot and cold composting handle bulk quantities of mixed materials. Vermicompost is nutritionally denser and biologically richer than most standard compost; it also requires maintaining a living organism population rather than managing a passive process.

Q18: What is bokashi and is it really composting?

Technically, bokashi is fermentation rather than composting — it preserves organic matter in an acidified state rather than decomposing it. The output, however, functions as a soil amendment in the same way that compost does, and when incorporated into soil or added to a traditional pile, it completes its decomposition within two to four weeks. Whether to call it composting is a definitional question;

whether it is useful is not — it handles inputs that aerobic composting cannot and fits living situations where traditional composting is impractical.

Q19: Can I compost in winter?

Cold composting slows dramatically below 50°F (10°C) and may appear to stop in freezing conditions, though activity continues slowly. Indoor vermicomposting continues normally as long as the bin is kept above 55°F (13°C). Bokashi is temperature-independent within normal indoor ranges. Hot composting can maintain internal temperatures above freezing even in cold climates if pile mass is sufficient and materials are added regularly; insulating the exterior with straw bales or a tarp extends the active season significantly.

Q20: How long does composting take?

Vermicomposting: two to three months for first harvest, then continuous. Hot composting: four to eight weeks for active phase, plus two to four weeks of curing. Cold composting: six months to two years. Bokashi: two weeks fermentation, then two to four weeks for soil incorporation. The variation across methods is wide enough that "how long does composting take" is more usefully answered by specifying the method.

Using Compost

Q21: How do I know when my compost is ready?

Three indicators together: it smells like earth, not like decomposing material; it has a dark, uniform color and crumbly texture with no recognizable original ingredients; and the bag test (sealing a sample in a plastic bag for a week) produces no ammonia smell. Any one indicator is suggestive; all three together confirm maturity. Rushing application of immature compost is one of the most common and consequential mistakes in home composting.

Q22: How do I use finished compost in my garden?

Top-dress established perennials and lawns at one to two inches per application; incorporate two to four inches into vegetable beds before planting; work a shovelful into planting holes for transplants; amend potting mixes at 20–30% by volume. See Book V Chapter 2 for the full application guide by crop type and method.

Q23: Can I use compost in potted plants?

Yes, as a component of a mix — not as the entire growing medium. Twenty to thirty percent compost blended with perlite, bark, or coir provides fertility without the drainage problems and anaerobic risk of pure compost in a container. Refresh by replacing or supplementing 20–30% of the mix each growing season.

Q24: What is compost tea and how do I make it?

Aerated Compost Tea (ACT) is made by aerating one part finished compost in five to ten parts water for 24–36 hours, using an aquarium pump and aeration stones. The aeration maintains aerobic conditions that allow beneficial bacteria to multiply. Add one to two tablespoons of unsulfured molasses per five gallons as a microbial food source. Strain and use within four hours of completion. See Book V Chapter 3 for the full brewing and application guide.

Q25: Can I use too much compost?

Rarely at normal application rates, but the risk is real at very high concentrations. Phosphorus accumulation from repeated heavy applications can lock out other nutrients and contribute to runoff. Organic matter percentages above 8–10% in a bed produce diminishing returns. The practical guideline: two to four inches incorporated annually is appropriate for vegetable beds; one to two inches for perennials; a quarter to half inch for lawns.

Urban Composting

Q26: How do I compost without outdoor space?

Vermicomposting for plant-based kitchen waste, bokashi for the complete food waste stream including meat and dairy, and electric countertop composters for households that prefer appliance management over biology. A detailed comparison of all three systems by space requirement, input type, management intensity, and output is in Book IV Chapter 2.

Q27: Will composting smell in my apartment?

A correctly managed worm bin smells like healthy garden soil. A sealed bokashi bin produces no detectable odor at the exterior between feedings. The smell problems that apartment composters experience almost always trace to the collection stage — food scraps in an open bowl or loosely covered container for several days — rather than from the composting system itself. A well-designed composting setup in a well-managed apartment should not be detectable by smell from across the room.

Q28: Where can I drop off compost if I can't process it at home?

Municipal curbside organics collection is available in a growing number of cities. Community gardens and urban farms frequently accept food scraps from local residents. Organizations like ShareWaste (sharewaste.com) and apps including NextDoor's composting features map local composters willing to accept scraps from neighbors. Farmers markets often host compost drop-off programs organized by local composting operations.

Sustainability

Q29: Does composting at home really make a difference?

Yes, in two ways. First, it diverts organic waste from landfills, reducing methane generation — approximately 0.5 to 1.0 metric tons of CO_2-equivalent per household per year, comparable to removing several hundred miles of annual car travel. Second, it improves soil health in ways that reduce the need for synthetic fertilizer and irrigation inputs, both of which carry their own environmental costs. The individual impact is real and measurable; it is not, by itself, a solution to system-scale problems, and the honest case for composting does not need it to be.

Q30: What's the single most impactful thing I can do to improve my composting results?

Commit to a specific method and run it consistently for one full year. The most common composting failure is not bad technique — it is abandonment after a slow start or a first problem. Every composting method has a learning curve, and the gardeners who produce the most consistently excellent compost are those who have accumulated enough seasons of practice to read their system well. The year of consistent effort is the investment; everything after that is the return.

Afterword

There is a corner of my garden that I have been composting for twenty years. The soil there is so dark it looks almost black, even in dry weather. It holds moisture through July. Earthworms surface without encouragement after rain. The plants I grow in it produce more, fail less, and recover faster from the things that go wrong than anything I grow anywhere else on the property.

I mention this not to make a grand claim about composting but to be specific about what two decades of returning organic matter to one place actually looks like. The improvement was not dramatic in any single year. It was barely perceptible in the first few. It accumulated in the way that most soil improvements accumulate — gradually, quietly, and then suddenly, one spring, you realize the bed has become something different from what it was.

That trajectory is the honest shape of composting's value. It does not rescue a garden in a season. It builds something over seasons that cannot be bought or shortcut — a living soil that works with you rather than against you, that responds to inputs rather than simply receiving them, that carries the accumulated biology of everything you have put back into it.

The science in this book is accurate to the best of my knowledge as of the time of writing, and I have tried to mark the boundaries between what the evidence supports and what remains contested. The practices are ones I have used myself or observed closely enough to recommend with confidence. Where I have expressed opinions — about which methods suit which situations, about what matters and what doesn't — they are grounded in experience, but they are still opinions, and your garden will have its own answers.

Start somewhere. One pile, one worm bin, one bokashi bucket. Manage it through one full year. By the end of that year, you will know more about your soil, your waste stream, and how decomposition actually works than any book can teach you in advance. The knowledge that comes from running a system through all four seasons is qualitatively different from the knowledge that comes from reading about it.

The compost will be ready long before you are done learning from it.

Thanks

Thank you for reading Composting Mastery. Writing this book has been a labor of genuine passion, and I hope it serves you well — in your garden, on your balcony, or wherever you choose to compost.

If this book has been useful to you, I'd be grateful if you could leave a review on Amazon. As an independent author, every review makes a real difference — it helps other readers find this book and helps me keep writing. It only takes a minute, and it means more than you might think.

— *Oliver Thorne*

Further Reading & References

Essential Books

- Howard, Albert. *An Agricultural Testament.* Oxford University Press, 1940.

- Appelhof, Mary. *Worms Eat My Garbage.* Updated ed., Storey Publishing, 2017.

- Lowenfels, Jeff, and Wayne Lewis. *Teaming with Microbes: The Organic Gardener's Guide to the Soil Food Web.* Timber Press, 2010.

- Lanza, Patricia. *Lasagna Gardening.* Rodale Press, 1998.

- Martin, Deborah L., and Grace Gershuny, eds. *The Rodale Book of Composting.* Rodale Press, 1992.

- Campbell, Stu. *Let It Rot! The Gardener's Guide to Composting.* Storey Publishing, 1990.

- Pleasant, Barbara, and Deborah L. Martin. *The Complete Compost Gardening Guide.* Storey Publishing, 2008.

Key Scientific & Institutional Sources

- Food and Agriculture Organization of the United Nations. *Food Wastage Footprint: Impacts on Natural Resources.* FAO, 2013.

- U.S. Environmental Protection Agency. *Composting At Home.* epa.gov/r ecycle/composting-home.

- U.S. Department of Agriculture, Natural Resources Conservation Service. *Soil Health*. nrcs.usda.gov.

- Higa, Teruo. *An Earth Saving Revolution*. Sunmark Publishing, 1991.

Organizations & Resources

- US Composting Council — compostingcouncil.org

- Rodale Institute — rodaleinstitute.org

- GrowNYC — grownyc.org

- Cornell Composting — compost.css.cornell.edu